Edwin M Bruce
Manual for Detection of the Common Food Adulterants

First Edition

By
IAPEN

About Edwin M Bruce

EDWIN M. BRUCE was an instructor in chemistry INDIANA STATE normal school , presently known as Indiana University of Pennsylvania, USA, March 25, 1907.

The present updated version is inspired by Edwin M Bruce.

Editorial Team
IAPEN Simple Protocols for Detection of Common Food Adulterants Project - 2012

Edited by

R R Siva Kiran
Assistant Professor, Department of Biotechnology, M.S. Ramaiah Institute of Technology, Bangalore, India

Team In-charge

VIGNESH.S
7th Semister, B. E (Biotechnology)
Department of Biotechnology, M.S. Ramaiah Institute of Technology (MSRIT), Bangalore, India

Team Members

VEENA VISWANATHAN, 7th Semester, B.E (Biotechnology), MSRIT
JEEVA LAKSHMI V, Research Associate, Department of Biotechnology, MSRIT
SHEELA K. P., Instructor, Department of Biotechnology, MSRIT
SHARADA B., Instructor, Department of Biotechnology, MSRIT
BHARATEESH R. S., 7th Semester, B.E (Biotechnology), MSRIT
SANGAMITHRA N., 7th Semester, B.E (Biotechnology), MSRIT
KAVYAA VENKAT, 5th Semester, B.E (Biotechnology), MSRIT
SURABHI BHAT, 5th Semester, B.E (Biotechnology), MSRIT
THULASI GOVARDHAN, 5th Semester, B.E (Biotechnology), MSRIT
VIBHA B., 5th Semester, B.E (Biotechnology), MSRIT
KEERTHI K. SHENOY, 3rd Semester, B.E (Biotechnology), MSRIT
YUGADHI. AJOY, 3rd Semester, B.E (Biotechnology), MSRIT

Cover page designed by ISWAKU RASTOGI, 7th Semester, B.E (Biotechnology), MSRIT, Bangalore

About IAPEN

The Indian Association for Parenteral and Enteral Nutrition (IAPEN) is an organization in the field of parenteral and enteral nutrition and promotes basic research, clinical research, advanced education, organization of consensus statements about clinical care and quality control.

IAPEN is the additional link to the already existing chain consisting of American Society for Parenteral and Enteral Nutrition (ASPEN), European (ESPEN, BAPEN), Australian (AusPEN), Parenteral and Enteral Nutrition Society of Asia (PENSA) and South African (SASPEN) societies.

Mission statement

To help ensure that those suffering from malnutrition or other nutritional problems are appropriately recognised and managed (adopted from BAPEN, UK)

To help ensure that proper nutrition is supplied to children in orphanages, anganwadis and schools, students in hostels and pg accommodations, patients in hospitals, senior citizens in old age homes and customers in restaurants.

Aim

To improve the nutritional care of people at risk of malnutrition and food adulteration whether in hospitals or in the community.

To improve the basic and advanced clinical research and to organize the consensus statements about clinical care and care quality control.

About IAPEN Simple Protocols for Detection of Common Food Adulterants Project - 2012

IAPEN strongly believes that food adulteration is one of the major cause of malnutrition in India. The food is adulterated in various forms before it reaches the children. Addition of stones in rice, milk in water, diluted rasam or sambar, soda in cooked rice, colors in turmeric and saw dust in chilli powder, trans fats in oils etc. Lack of testing laboratories, food experts, lack of training to the food safety officers, lack of equipments and infrastructure etc. implementing the Prevention of Food Adulteration Act by Government of India is becoming difficult.

IAPEN has come up with a simple and an economical solution for fighting against the food adulteration and for strict implementation of "Prevention of Food Adulteration Act, 1956", by popularizing simple protocols for detection of food adulteration. The protocols can be executed with the help of chemicals present in Pre-University Colleges (PUC) or schools. New protocols will be added to this book every year after scrutinizing in the annual meetings of IAPEN by expert members.

RNL Naidu
Hon Secretary
IAPEN

Preface

Because of the recent agitation of the pure food question throughout the country, health officers and food-inspectors, are constantly called upon to test the purity of various foods. And this usually involves nothing more than making simple qualitative tests for adulterants. In view of the fact that there is now no text or manual devoted exclusively to the qualitative examination of foods, this little book is offered to those who are interested in this work.

Its aim is to bring together in one small book the best and simplest qualitative tests for all the common food adulterants. It contains a brief statement of the adulterants likely to be found and the reason for their use. It is hoped that it will be specially valuable to food safety labs in furnishing excellent supplementary work in qualitative analysis. But it is hoped that it will find its greatest usefulness in contributing something toward the great pure food reform.

The updated version includes 210 simple protocols collected from The Joint FAO/WHO Expert Committee on Food Additives (JECFA), Food and Agriculture Organization of the United Nation Organization, 29 protocols from The Directorate of Prevention and Food Adulteration, Government of Delhi, India and Directorate of Marketing & Inspection (DMI), Ministry of Agriculture, Government of India and 129 protocols from the "Quick test for some adulterants in foods" an instruction manual developed by Food Safety and Standards Authority of India, Government of India.

Editorial Team

The Indian Association
for Parenteral and Enteral Nutrition

Contents

(Vegetable oil Ghee Honey Pulses/ Besan, Pulses Bajra, Wheat flour Common spices like Turmeric, chilly, curry powder etc. Black Pepper, Spices (Ground), Coriander powder Chillies, Badi, Elaichi seeds, Turmeric Powder, Turmeric, Cumin seeds (Black jeera), Asafoetida Heeng), Food grains)

(Milk, Sweet Curd, Rabdi, Khoa and its products (51), Chhana or Paneer, Ghee, cottage cheese, condensed milk, khoa, milk powder etc, Ghee, Butter, Oils and fats (52), Mustard oil, Edible oil, Coconut oil (54), Sugar, Pithi Sugar, Honey, Jaggery, Honey, Bura sugar, Sweetmeats, ce-cream and beverages, Wheat, Rice, Maize, Jawar (55), Bajra, channa, Barley (56) etc., Sella Rice (Parboiled Rice), Parched rice (57), Maida/ Rice, Maida, Food grains, Wheat flour, Wheat flour, Dal whole and spilt (58), Atta, Maida, Suji (Rawa), Bajra, Sago (59), Besan, Pulses, Whole spices, Black pepper (60), Cloves, Mustard seed, Powdered spices (61), Turmeric powder, Turmeric ....

Contents

Contents

LIST OF FOOD ADULTERANTS AND FOOD ADDITIVES AND SIMPLE ANALYTICAL PROCEDURES

Dairy products
Milk

Formaldehyde

Hehner's Sulfuric Acid Test: Put 10 cc. of the suspected milk in a wide test tube and pour carefully down the side of the inclined tube about 5 cc. commercial sulfuric acid so that it forms a separate layer at the bottom. A violet coloration at the union of the two liquids indicates the presence of formaldehyde. If the commercial acid is not available, the pure acid may be used, but a few drops of ferric chloride must be added.

Hydrochloric Acid Test: 2 cc. of 10 per cent ferric chloride is added to one liter of commercial hydrochloric acid sp. gr. i. 2 (or any quantity in this proportion). To 10 cc. of this mixture add 10 cc. of the milk to be tested. Heat the mixture slowly nearly to the boiling point, in an evaporating dish, but agitating it all the while to prevent the curd collecting in one mass. If formaldehyde is present, there will be a violet coloration. It is said that by this test as small a quantity of formaldehyde as i part in 250,000 parts of milk can be detected. It is not so sensitive in sour milk.

Boric acid

Turmeric Paper Test: Incinerate some of the milk and acidulate the ash with a very few drops of dilute hydrochloric acid and afterwards dissolve it in a few drops of water. Place a strip of turmeric paper in this solution for a few minutes, then remove and dry it. If boric acid either free or combined is present, the turmeric paper will be turned to a cherry-red color. (Turmeric paper may be prepared by dipping pieces of smooth, thin filter paper in a solution of powdered turmeric in alcohol.).

Another way of testing: Make strongly alkaline with lime water, 25 grams of the milk, and evaporate to dryness on the water bath. Destroy the organic matter by igniting the residue. Dilute with 15 cc. of water and acidify with hydrochloric acid. Then add i cc. of the concentrated acid. Dip a piece of delicate turmeric paper in the solution; and if borax or boric acid is present, it will have a characteristic red color when dry. Ammonia changes it to a dark blue green, but the acid will restore the color.

Salicylic acid   Dissolve one gram of mercury in 2 grams of nitric acid (sp. gr. 1.42) and then add to the solution the same volume of water. Add 1 cc. of this reagent to 50 cc. of the milk to be tested, and shake and filter. The perfectly clear filtrate is shaken with ether and the ether extract evaporated to dryness. Then add a drop of ferric chloride solution, and a violet color will be produced if salicylic acid is present.

Annatto   Add acid sodium carbonate to a sample of the milk until it shows a slight alkaline reaction. Immerse a piece of filter-paper and leave it in for 12 or 15 hours. If annatto is present; there will be a reddish-yellow stain on the paper.

Caramel   **Leaches Method** — Warm 150 cc. of the sample and add 5 cc. of acetic acid, then continue heating it nearly to the boiling point, stirring while it is being heated. Separate the curd by gathering it with the stirring rod or by pouring through a sieve. Press out all the whey from the curd and macerate the latter for several hours (12 hours) in 50 cc. of ether. It is best to do this in a tightly corked flask, shaking it frequently. If the milk was uncolored or colored with annatto, the curd when thus treated will be white. If the curd is a dull brown color caramel was probably used to color the milk. Confirm its presence by shaking a portion of the curd with concentrated hydrochloric acid (sp. gr. 1.20) and gently heating. If the acid solution turns blue while the curd does not change its color, caramel was used to color the milk.

Coal-tar dye
Lythgoe's Method — Mix in a porcelain vessel about 15 cc. each of the sample of milk and hydrochloric acid (sp. gr. 1.20) and break up the curd into coarse lumps by shaking gently. If an azo-color was used to color the milk this curd will be pink, but the curd of normal milk will be white or yellowish.

Gelatin
A. W. Stokes Method — Dissolve 1 part by weight of mercury in 2 parts of nitric acid (sp. gr. 1.42). Add 24 times this volume of water. Mix equal volumes (about 10 cc.) of this reagent and the milk or cream, shake well and add 20 cc. of water. Shake again and, after standing 5 minutes, filter. When a great quantity of gelatin is present, the filtrate will be opalescent instead of perfectly clear. To a little of this filtrate in a test tube add the same volume of a saturated aqueous solution of picric acid. If much gelatin is present a yellow precipitate is produced, smaller amounts produce cloudiness. If the filtrate is perfectly clear, gelatin is absent and picric acid may be added without producing any noticeable effect.

Starch
The presence of starch in milk may be detected by heating a small quantity of the milk to boiling. When it has cooled add a drop of iodine in potassium iodide, and if starch is present there will be a blue coloration.

Butter

Coloring Matter
Martinis Test: Add 2 parts of carbon bisulfid, a little at a time and with frequent shaking, to 15 parts of alcohol. Shake 25 cc. of this solution with 5 grams of the butter, and let stand for some

time. The carbon bisulfide dissolves out the fatty matter and settles to the bottom. The alcohol remains on top and will dissolve out any artificial colors that may be present. If only a little coloring matter is present use more of the butter.

Annatto — Evaporate a portion of the extract to dryness and add sulfuric acid to the residue. If annatto is present a greenish-blue color forms. Should a pink tint result the presence of a coal-tar color is to be suspected.

Saffron — When saffron is present, nitric acid colors the alcoholic extract green, and hydrochloric acid colors it red.

Turmeric — Add ammonia to the alcoholic extract, and if it turns brown it indicates the presence of turmeric.

Marigold — Add silver nitrate to the extract, and if it turns black the presence of marigold is indicated.

Process or Renovated Butter — Heat a little of the suspected butter in a spoon or dish, and if it is process butter it will sputter, but not foam much. Make the test also with some butter known to be pure and fresh.

Coal-tar — *Geisler's Method:* To a few drops of the clarified fat on a porcelain surface, add very little fullers' earth. If a pink to violet-red coloration is produced in a short time the presence of an azo-color is indicated.

Oleomargarine — Immerse a test tube, containing some of the filtered fat, in boiling water for 2 minutes. Make a mixture of 1 part glacial acetic acid, 6 parts ether, and 4 parts alcohol. Add to 20 cc. of this mixture in a 50 cc. test tube, 1 cc. of the heated fat which may be transferred by means of a hot pipette. Stopper the tube and shake it well. Immerse in water at 15° or 16° C. Pure butter when thus treated remains clear for quite a while. There will be only a very little deposit after standing an hour, but oleomargarine gives a deposit almost immediately, and in a few minutes there will be a copious precipitate. When the oleomargarine in butter is in about the proportion of 1:10, it will not separate much short of 15 minutes.

Pure butter test   Hess and Doolittle Test. — Melt some of the butter (say 40 grams) at about 50° C. If the butter is pure and fresh the melted fat will clear up almost as soon as it is melted, while the fat of process butter remains turbid for quite a while. After most of the curd has settled, decant as much as possible of the fat. Pour the remainder on a wet filter, and boil. If it was ordinary butter this filtrate will become milky, but if process butter a flocculent precipitate will form.

Meat

Potassium Nitrate (Saltpeter)   Corned and smoked meats are usually preserved with saltpeter. Since smoked and cured meats are used in making potted meats, saltpeter is quite frequently found in the latter. It may be detected by the usual test for nitrates since no other nitrate is apt to be present, though one may identify the metal by the qualitative test for potassium. To test for nitrates treat a little of the meat with 2 or 3 cc. of a 1 per cent solution of diphenylamine in strong sulfuric acid. If a nitrate is present, a deep blue color forms instantly, which is not obscured by the charring effect of the acid.

Boric Acid   Pick apart into fine pieces about 25 or 50 grams of the lean meat and warm with a little water which has a few drops of hydrochloric acid in it. Soak a piece of turmeric paper in the extract, and if boric acid is present the paper will be colored rose-red when it is dry. A weak alkali turns this colored paper olive. Another method is to burn a piece of the meat to an ash, after being treated with lime water. Make a solution of the ash and make slightly acid with hydrochloric acid. Then test with the turmeric paper with the same results.

Sulfurous Acid   Digest 40 or 50 grams of the meat in hot water, treat with 10 cc. glacial phosphoric acid to coagulate the proteins. Strain through a cotton bag and transfer the filtrate to a short-necked flask and distil receiving the first part of the distillate in a solution of iodine. Boil, and add barium chloride. If sulfurous acid is present, it will be oxidized to sulfuric acid and precipitated as barium sulfate by the barium chloride. More than a mere trace of the precipitate proves that some sulfite was used to preserve the meat. Another method is to place the meat on paper, which has been saturated with potassium iodate moistened with dilute sulfuric acid (1:8); nitric oxide must not be present. If sulfurous acid is present a deep blue color forms at once. A trace of this color may form after some time with meat that is not fresh; hence this method cannot be used in examining canned meat.

Salicylic Acid   Heat 50 grams of the meat in 50 cc. of water. Add 10 cc. of a strong solution of glacial phosphoric acid and strain through a cotton bag. Extract the filtrate with a little ether (about 50 cc.) in a separating funnel. Let the ether evaporate spontaneously. Take up the residue with 3 cc. of water, and add one or two drops of a one-half per cent solution of ferric chloride. If salicylic acid is present the mixture will be purple or violet. The same test is performed by slightly acidifying a portion of the lean meat; then extracting with ether, and evaporating to dryness and testing the residue with a drop of ferric chloride solution. A deep violet coloration is produced if salicylic acid is present.

Benzoic Acid

Mohler's Method: Prepare a sample as in the test for salicylic acid by heating 50 grams of the meat in 50 cc. of water. Add 10 cc. of a concentrated solution of glacial phosphoric acid, and strain through a cotton bag. Neutralize with sodium hydrate and evaporate to dryness or to a small volume. After treating with 3 cc. of concentrated sulfuric acid, heat till white fumes appear. Add 4 or 5 crystals of potassium nitrate and continue heating until the solution is colorless or nearly so. When cool dilute with water, add an excess of ammonia, and place in a narrow vessel like a test tube. Add one or two drops of ammonium sulfide carefully so that the liquids do not mix. If the surface of the liquid immediately becomes red, benzoic acid is present. If this test is not carefully performed, it is worthless, as other substances give similar results. Confirm its presence by neutralizing the aqueous solution of the extracted benzoic acid with sodium hydroxide; concentrated to a very small volume. Acidify with sulfuric acid. A white flocculent precipitate shows the presence of considerable benzoic acid.

Diseased Meat

The following method is recommended by Ebsers: Hold a small piece of the suspected meat over a mixture of i cc. hydrochloric acid, 3 cc. alcohol, and i cc. of ether. The formation of ammonium chloride fumes shows that decomposition has begun. Do not mistake the fumes of the acid for those of ammonium chloride.

Canned meat

Heavy Metals

A. H. Allen's Method: About 25 grams of the substance is mixed slowly with enough strong, pure sulfuric acid to just moisten the mass, avoiding an excess. Heat on a water-bath for a short time, then raise the temperature gradually, and maintain till the chlorides seem to be decomposed. It must not be hot enough, however, to volatilize the sulfuric acid. Then add 1 cc. of strong nitric acid and heat till red fumes are given off. Freshly ignited magnesia in the proportion of 0.5 gram for each cc. of sulfuric and nitric acid previously used is now stirred into the mass and the whole ignited at a dull red heat. This is best done in a gas-muffle. When cool, moisten the ash with nitric acid and gently re -ignite, repeating this treatment till the carbon is entirely consumed. Treat the residue with 8 or 10 drop of strong sulfuric acid, heat till fumes are given off, cool, boil with water, dilute to about 100 cc. and saturate with hydrogen sulfide, filter, examine as follows:

Zinc and iron may be in solution. Add bromine water to destroy hydrogen sulfide and to oxidize the iron, boil and add ammonium hydrate in excess, boil again and filter. The precipitate will contain the iron and the phosphates. Filtrate, when blue, proves the presence of nickel. Heat to boiling and add potassium ferrocyanide. A white precipitate or turbidity indicates zinc.

Lead, tin, copper, and calcium, if present, will be in the precipitate and residue. Fuse the mass in a porcelain crucible for at least ten minutes with 2 grams each of potassium and sodium carbonates and half as much sulfur. After cooling, boil with water and filter. To the residue, add hydrochloric acid and boil as long as hydrogen sulfide is given off. A few drops of bromine water will complete the oxidation of the copper sulfide. To the filtrate, add ammonium hydroxide in excess. When the filtrate is blue, it indicates the presence of copper. Acidify the filtrate with acetic acid and test for lead by adding potassium chromate, a yellow precipitate being formed when it is present. The filtrate may contain tin. If tin is present, a yellow precipitate of stannic sulfide will be formed.

Coloring Matter	Sausages and other chopped meat preparations, together with corned meat that has been cured without saltpeter, are often treated with artificial coloring matter. Aniline red and cochineal carmine are usually employed for this purpose. The former may be detected by picking the meat

apart and treating it with methylated spirit, strain or filter the extract and take up with water. Then a piece of white wool is immersed in the boiling liquid and, if it is dyed red, rosaniline is present. Cochineal-carmine may be as follows: Cut up fine about 20 grams of the meat and heat in a water-bath with water and glycerine mixed in equal parts. If the above coloring matter is present the liquid will become quite red in color, if not present a slight yellow color results from this treatment. If a spectroscope is available this dye is easily recognized.

Starch

In Sausage, Deviled Meat, and Similar Products: Cracker and bread crumbs are often added to these preparations and their presence is best detected by examining the amount of starch present. Do this by boiling some of the sample in water, and when cool adding a drop or two of Iodine reagent. The usual blue color is produced if much starch is present. If there is only a little starch, it may be necessary to examine it under the microscope to determine whether the starch is from the pepper and other spices used or from some cereal.

Horse Flesh in Sausage and Mince Meat

Horse flesh is detected by testing for glycogen, which is present in it in larger quantities than in other meats

Courley Coremon's Test: Boil 50 grams of the meat for a half hour with water, strain, and to a portion of the filtrate add a few drops of potassium iodide-iodine solution (potassium iodide 0.4 gram; iodine 0.1 gram; water 20 cc). If considerable horse meat is present the glycogen will color the liquid dark brown, which disappears on heating, but returns on cooling.

Eggs

Delarne's Test: Place the egg in a 10 per cent solution of common salt. Perfectly fresh eggs sink to the bottom. Those remaining immersed, but suspended in the liquid, are at least three days old, while those rising to the surface and floating are more than five days old. The older the egg, the higher it floats and the more it will stand on end. This test is not applicable to eggs that have been preserved. Hold the egg between a bright light and the eye, and if the air chamber is small, and no dark spots but instead a rather uniform rose-colored tint is seen, the egg is fresh. If the contents appear cloudy and the air chamber larger, the egg is not fresh. The darker the contents of the egg, the older it is.

Cereal Products (Flour)

Alum

Wynther Blyth Method: Add a little water to the sample and macerate. Soak pieces of gelatin in the solution and leave for a half day, remove the gelatin and dip the pieces in a mixture of equal volumes of a fresh tincture of logwood and a saturated solution of ammonium carbonate. The gelatin strips will turn blue if alum is present.

Bell & Carter Method: Make a fresh 5 per cent tinc ture of logwood in methyl alcohol. Dampen about 10 grams of the flour with water and add i cc. of the logwood tincture and the same quantity of a saturated solution of ammonium carbonate. Pure flour gives a pinkish color which fades to buff or brown. The presence of alum produces a lavender or bluish tint which becomes more distinct as it dries.

Copper sulfate

This adulterant may be detected in either flour or bread, by soaking the flour or bread in a dilute solution of potassium ferrocyanid acidulated with acetic acid. If copper be present a purplish or reddish-brown coloration will be produced.

Substituted flours

Voge Vs Method: Make a mixture of alcohol (70 per cent), 95 parts, hydrochloric acid 5 parts. Treat a sample of the flour in a test tube with this reagent. Shake well. Heat to boiling and allow to settle. A colorless fluid shows the flour to be pure, a straw-colored tint indicates the presence of gruffs with bran, an orange-yellow proves the presence of corn-cockle flour, a flesh-colored liquid indicates the presence of ergot, while a green color indicates buckwheat flour.

Corn meal in wheat flour

Kraemer claims to be able to detect as small amount as 5 per cent of maize in wheat flour, by the following test. Mix a gram of the flour with 15 cc. of good glycerin, and heat to boiling for a short time. If corn meal is present, there will be an odor like that of pop corn.

Wheat in rye flour

Kleeburg recommends the following test: A little of the flour is mixed on a piece of common window glass or microscope slide, with sufficient water (at 45°C.) to float the flour particles. Spread the mixture out over the glass, and press another glass down upon it. When wheat flour is present, white spots will be seen, and if the glasses are slid upon each other the spots will pull out into threads, and the thicker and longer they are the more wheat flour there is present.

Ergot in rye flour

Boettger gives the following chemical test for ergot: Heat 10 to 15 minutes with an equal quantity of ether, adding a few crystals of oxalic acid. When ergot is present a reddish color develops. Another Method. — Bui. 51, Bureau of Chem. Digest 20 grams of the suspected flour, Vvith boiling alcohol, till no more color is extracted. Add i cc. of sulfuric acid (i: 3), and if ergot is present the solution will be colored red.

Bread

Alum

Moisten a piece of the bread with water, and then with a logwood solution (5 grams logwood digested in 100 cc. of alcohol). If alum is present the bread will become lavender blue in two or three hours. Pure bread would have a red-brown tint. To prove the presence of alum, the blue color must be permanent at the temperature of boiling water. (The logwood used in this test must be pure.)

Blyth's Test: Macerate 150 grams of the sample for 45 or 50 hours in a couple liters of water; after straining through muslin, evaporate to a small volume over a low flame. Immerse a strip of gelatin in this liquid, and then in a logwood solution (same as in last test), and if alum is present it will acquire the lavender color. If the bread in either of these tests is sour, the following modification (Vanderplanken) must be made. Reduce 15 grams of the sample to a paste with water and some pure chloride of sodium, adding 10 drops of a fresh alcoholic solution of logwood, after which add 5 grams of pure potassium carbonate. Mix well, and after washing with 100 cc. of water into a vessel allow to settle. If alum is present the liquid will soon become reddish-violet, and if not present it will be blue.

Copper Sulfate   This adulterant may be detected in either flour or bread, by soaking the flour or bread in a dilute solution of potassium ferrocyanid acidulated with acetic acid. If copper be present a purplish or reddish-brown coloration will be produced.

Ginger cake   Tin may be detected by the method for heavy metals under meat.

Baking powder

Tartaric acid   **Wolff's Method:** If no starch is present, mix a little of the powder with some dry resorcin. Add a few drops of sulfuric acid and heat gently. A rose-red color forms if tartaric acid or tartrates are present. The color should disappear when diluted with water. When starch is present, mix well by shaking about 5 grams of the powder with 250 cc. of cold water. Let the insoluble matter settle and pour the liquid upon a filter. Evaporate the filtrate to dryness, treat the powdered residue with a 1 per cent solution of resorcin. Add 3 cc. of strong sulfuric acid, heat slowly. A rose-red color forms if tartaric acid is present. The color should be destroyed on the addition of

Tartaric acid Free	water. This test is applicable in the presence of phosphates and the acid may be free or combined. Make an absolute alcoholic extract of 5 grams of the powder and evaporate the alcohol. Add sufficient dilute ammonia to dissolve the residue, place in a test tube and drop in a crystal or two of silver nitrate. Heat gently, and a silver mirror will form if tartaric acid is present.
Sulfates; calcium, etc.	Boil a portion of the sample gently with strong hydrochloric acid, add barium chlorid. A white precipitate of barium sulfate will form if sulfuric acid is present.
Gypsum; calcium sulfate	Ash a portion of the sample and make the usual qualitative tests for calcium sulfate.
Ammonium salts	Extract a few grams of the sample with cold water, boil the extract with sodium hydroxid and place a piece of moist red litmus paper in the steam. It will be colored blue if ammonia is present.
Alum	Reduce to an ash about 2 grams of the powder in a platinum dish. Extract with boiling water, add ammonium chloride solution to the filtrate until a distinct odor of ammonia is given off. If a flocculent precipitate forms it indicates the presence of alum. This test for alum is applicable in the presence of phosphates. Mrs. Richards. Cover some logwood chips (they must be pure) with water and bring to a boil. Repeat this four times, saving only the last decoction. Shake some of the sample (a couple of teaspoonfuls) in a beaker half full of water. When it ceases effervescing, strongly acidify with acetic acid. Add a few drops of the logwood extract, and if alum is present a bluish-red color will appear.
Cream of tartar	Cream of tartar is bitartrate of potassium and is obtained from the lees deposited in wine casks. If gypsum has been used to clarify the wine, it will be present in the cream of tartar as calcium tartrate. Other adulterants of cream of tartar are acid calcium phosphate, starch, gypsum, and

alum.

Free or combined tartaric acid	If the sample is known to be free from starch the following test may be made: Mix a bit of the powder with a small quantity of dryresorcin and add a few drops of concentrated sulfuric acid. Heat slowly, and if a rose-red color forms, which disappears when diluted with water, there is present either tartaric acid or a tartrate. Then the sample contains starch, shake about 4 or 5 grams of it a number of times with 250 cc. of cold water in a large flask. Pour on a filter after the insoluble material has settled and evaporate the filtrate to dryness. The residue is to be tested for tartaric acid and tartrates, the same as when starch was absent.
Aluminium salts	Mix equal quantities (about 1 gram) of the sample and sodium carbonate and burn to an ash. Extract with boiling water and filter. Add to this filtrate enough ammonium chloride solution to cause a distinct evolution of ammonia. The formation of a flocculent precipitate shows the presence of aluminum. This test may be used when phosphates are present.
Ammonia Carbonate	Present in the Form of Ammonia Alum or Ammonium. Make a cold water extract of the powder and boil it with sodium hydroxide. Test the steam with moist red litmus paper.
Earthy materials	Treat the sample with warm potassium hydroxide. A residue indicates some earthy material.

Canned and bottled vegetables: preservatives: It is best to make a systematic examination for the different preservatives. The sample may be prepared by mixing 50 grams of the pulped material with sufficient water in a 250 cc. graduated flask. Add phosphoric acid till distinctly acid in reaction. Fill to the mark with water. Place in a distilling flask, and distil in a linseed oil or a paraffin bath till 30 cc. have been collected. Save this distillate for the following tests.

Formaldehyde

To 5 cc. of the above distillate in a test tube, add 2 or 3 drops of a 1 per cent aqueous solution of phenol and mix well. Incline the tube and carefully pour down the side 5 cc. of concentrated commercial sulfuric acid so that the two liquids do not mix. If formaldehyde is present there will be a crimson zone at the plane of union of the solutions. This coloration takes place when the formaldehyde is present in the proportion of i part in 100,000 parts. When there is a greater quantity of formaldehyde present a white turbidity or a light-colored precipitate forms above the coloring.

Phenylhydrazine Hydrochloric Test: Dissolve 2 grams of phenylhydrazine hydrochloride and 3 grams of sodium acetate in 20 cc. of water. Add 2 to 4 drops of this reagent and the same number of drops of sulfuric acid to i or 2 cc. of the above distillate, to be examined in a test tube. A green coloration is produced when formaldehyde is present.

Hydrochloric Acid Test: Add 5 cc. of the distillate to be tested to about 5 cc. of milk known to be pure, and about 10 cc. of concentrated hydrochloric acid (sp. gr. 1.2) which contains i cc. of a 10 per cent ferric chloride solution to each 500 cc. of the acid. Heat slowly to 80° or 90° C. over the free flame, agitating it at the same time to break up the curd. A violet coloration indicates formaldehyde.

Sulfurous Acid and Sulfites

Mix 150 grams of the finely ground sample with enough water to make a thin paste. Acidify with phosphoric acid and distil till 25 cc. have been collected (The delivery tube of the condenser should dip below the surface of a little water). Treat the distillate with a few drops of bromine water and boil for a short time. If a precipitate forms on the addition of barium chloride, the presence of sulfurous acid is indicated.

Salicylic Acid   Acidify 50 cc. of the sample with sulfuric acid, and shake vigorously with 50 cc. of a mixture of equal parts of ether and petroleum spirit. When the liquids have separated, draw off as much as possible of the solvent and filter. If an emulsion forms use a centrifugal machine, and evaporate with a small flame. If needle shaped crystals form, salicylic acid is present. Add a few drops of water and a drop of very dilute ferric chloride solution in such a way that the solutions will come together slowly. The presence of salicylic acid gives a purple or violet color.

Saccharine   This is used quite extensively as a sweetening agent in canned sweet corn, and other similar products. Macerate about 20 grams of the sample after mixing with 30 to 40 cc. of water and strain through muslin. Acidify with 1 or 2 cc. of sulfuric acid (1 to 3) and extract with ether. (If an emulsion forms, use a centrifugal machine.) Separate the ether layer and let the ether evaporate spontaneously and use the residue in the following tests: Take up a part of the residue with water and taste. If it is very sweet saccharin is present. Confirm by the following:

Schmidt's Test: Add about 1 gram of sodium hydroxide to another part of the residue, and heat in an air-oven or oil bath, for half an hour at about 250° C, to convert the saccharin into salicylic acid. After it has cooled, acidify with sulfuric acid, extract and test for salicylic acid with 2 or 3 drops of ferric chloride solution, letting the solutions come together slowly. A purple or violet coloration proves the presence of salicylic acid, which in turn indicates the presence of saccharin. This test cannot be used if salicylic acid was used as a preservative in the original product. A test for the acid should first be made.

Bornsteiff's Test: Heat the remainder of the above ether residue with resorcin and a very little sulfuric acid till it begins to swell. (It is best to do this heating in a test tube.) Let cool till the action stops, heat again and repeat the operation several times. After cooling the last time, dilute with water and add sodium hydrate till neutral. If saccharin is present, there will be a red-green fluorescence.

Benzoic Acid

Acidify 50 cc. of the sample with sulfuric acid and shake vigorously with 50 cc. of a mixture of equal parts of ether and petroleum spirit. Let the liquids separate, and then draw off as much as possible of the solvent and filter. (Use a centrifugal machine if an emulsion forms) Separate the extract into 2 parts and evaporate each to dryness over a small flame and make the following test:

Ferric Chloride Test: Dissolve one of these residues in ammonia, and evaporate to dryness on a water-bath. Take up the residue with warm water, filter, and collect the filtrate in a small test tube. Add a drop of ferric chloride solution, and if benzoic acid is present a characteristic flesh or brownish colored precipitate of ferric benzoate forms. Sometimes in such products as sweet pickles, a basic ferric acetate precipitate comes down and the following test had better be applied.

Peter's Method: Take about 0.i gram of the second part of the above ether residue, place in a large test tube (about 50 cc.) and dissolve in 5 to 8 cc. of concentrated sulfuric acid. Add from 0.5 to 0.8 gram of barium peroxide, a little at a time. Shake each time and cool in water if necessary. This should produce a permanent froth on the sulfuric acid. Let stand 25 or 30 minutes, then fill the tube three fourths full of water, shake and cool rapidly to the temperature of the room, and filter off the barium sulfate. Extract with chloroform or ether. Remove the extract and test it for salicylic acid with dilute ferric chloride. (See first test under salicylic acid.) In this method salicylic acid must first be proven absent.

Mohler's Test: Treat the remainder of the second part of the above ether residue with 2 or 3 cc.

of concentrated sulfuric acid. Heat till white fumes appear. Add a few crystals of potassium nitrate and when cool dilute with water. Add an excess of ammonia, then a drop or two of ammonium sulfide. If a red color appears immediately on the surface, it shows the presence of benzoic acid.

Coloring matter in catsups and tomatoes

Cochineal

Girard and Dupre Test: Shake well a portion of the sample with water and filter, acidify with hydrochloric acid, then extract with amyl alcohol, and if cochineal is present the extract will be colored yellow or orange, the particular shade depending on the amount of cochineal present. Remove the amyl alcohol and wash with water until it is neutral. To half of this, add a very dilute solution of uranium acetate, drop by drop, and shaking well after the addition of each drop. Cochineal, if present, will produce a characteristic emerald-green color. Confirm by adding a drop or two of ammonia to the second half of the amyl alcohol extract and a violet coloration will be produced if cochineal is present.

Coal-Tar Coloring Matter

Sostegni and Carpentieri Test: Free from grease a piece of woolen cloth by boiling first in very dilute caustic soda solution and then in water. Acidify a portion of the sample with 2 to 4 cc. of 10 per cent solution of hydrochloric acid and filter. Strips of the cleansed cloth are boiled in this filtrate for 5 or 10 minutes, then removed, washed in water and boiled with very dilute hydrochloric acid solution. Wash out the acid and dissolve the color from the cloth by boiling in a solution of ammonium hydroxide (1 to 50) (The time required will depend upon the dye present). Remove the cloth from the solution and acidify the latter with hydrochloric acid and another piece of the cleansed cloth is immersed and again boiled. This second dyeing fixes only coal-tar colors on the cloth, hence, no fear of mistaking them for the natural color of the vegetable.

In green pickles, beans, peas, etc
Copper salts

Burn 20 grams of the sample to an ash and wet the ash with concentrated nitric acid, dilute with water and boil. Add ammonia till strongly alkaline and filter. If the filtrate is blue, copper is present. Confirm by acidifying the filtrate with acetic acid and adding potassium ferrocyanide. A red or brownish precipitate or coloration proves the presence of copper. The test for other heavy metals may be made by the general method given under meats.

In mixed pickles
Turmeric

Shake with alcohol to extract the color. Soak a piece of filter paper in the extract and dry in an air oven at 100° C. Wet the filter paper with a weak solution of boric acid to which a very little hydrochloric acid has been added. If turmeric is present, a cherry-red color will appear when the filter paper is dry.

"Soaked" Vegetables
Peas, Beans, and Corn

There is really no chemical test for this class of foods. All or nearly all of the green color of peas and beans is destroyed by the process of "soaking." They have the appearance of the well-matured product, and are firm and mealy with well-formed cotyledons. The process of soaking starts the growth of the caulicle of the pea. The kernel of corn is plump and hard and does not have the milky consistency of the immature product. The characteristic succulence of the green pea, bean, and corn is absent in the soaked product.

Alum in Pickles

This is sometimes added to the pickling solution to produce hardness and crispness. Burn to ash a sample of the pickles, and, if they are free from copper, fuse in a platinum dish with sodium carbonate. Extract with boiling water, and after filtering add ammonium chloride solution. If alum is present, a flocculent precipitate will form.

Examination of the 'Can' or 'Box' in which vegetables are sealed

Generally when the ends of a can are convex, instead of plane or concave, it is spoiled. In the souring of canned sweet corn, it is exceptional that the ends are forced outward. Strike the can and the spoiled cans will give a dull sound while the good ones will give a distinct tone. Some practice will be necessary to use this test. One can judge of the amount of tin dissolved by the corrosion of the inside of the can. Reject cans that show much rust around the cap on the inside of the head. If more than one hole is found soldered in the cap, reject the can. Cans of salmon are the only exception that hascome to the author's notice. A second hole, in general, indicates that decomposition had set in and the can had been punctured and resealed.

Fruits and fruit products (Preservatives)

Salicylic Acid

Place a few drops of this extract in a test tube and add a drop or two of a 0.5 per cent solution of ferric chloride. If salicylic acid is present, there will be a purple coloration.

Benzoic Acid

Mohler's Test: Add 2 to 3 cc. of strong sulfuric acid to a second portion of the above ether extract and heat until white fumes appear. Then add a few crystals of potassium nitrate and heat again. Continue adding the nitrate and heating till the solution is colorless or only a very light yellow. Dilute with about 5 cc. of water when cool, neutralize with ammonia. It should be filtered when not clear or when crystals of ammonium or potassium sulfate are formed. Add a few drops of ammonium sulfide to the filtrate in such a way as to prevent the mixing of the liquids. The sulfide will be on top. If a bright cherry-red color forms where the two liquids meet, either benzoic acid or saccharin is present. Distil and the benzoic acid will pass over, extract the distillate in the usual way and apply the above test to it for benzoic acid.

Saccharin

Taste a third portion of the ether extract. A very sweet taste indicates saccharin. A further test can be made by adding i or 2 grams of sodium hydroxide to the rest of the ether extract and heating a half hour in an oil bath at 250° C. Dissolve in water when cool, acidify with dilute sulfuric acid and extract with ether. The saccharin will have been converted into salicylic acid, which may be identified by the usual test for that acid. This test presupposes the absence of salicylic acid in the original material.

Fruits and fruit products (Colouring matter)

Coal-Tar Dyes

To attempt to identify the particular dye used in every case would be quite beyond the object of this set of simple tests. A general test showing the presence of a coal-tar dye is probably all that is usually desired.

Sostegni and Carpentieri Test: Such a test may be made by dissolving 15 grams of the fruit product in 100 cc. of water, filtering and acidifying with a small quantity of a 10 per cent solution of hydrochloric acid and again filtering. Place in the filtrate strips of white woolen cloth (nun's veiling will do) which have been freed from grease by boiling first in very dilute caustic soda solution, then in water, and boil for 5 to 10 minutes. Remove the cloth and wash it in water, then boil in very dilute hydrochloric acid. Stir the cloth in water to remove the acid and dissolve the color by boiling in a solution of ammonium hydroxid (i to 50). The time required will depend upon the particular dye used. Remove the cloth from the solution and acidify the latter with hydrochloric acid, a slight excess is better, and another piece of the cleansed cloth is immersed and again boiled. Nothing but coal-tar dyes will color in this second dyeing.

Cochineal

Girard and Dupre Test. — See tests for cochineal under "Catsups and tomatoes."

Acid Magenta

Girard and Dupre: Make about 100 cc. solution of the fruit, filter, and neutralize with potassium hydroxid (strength 5 to 100); about 2 cc. will be needed. Add 4 cc. of mercuric acetate solution (1 to 10), shake and filter. By this treatment the filtrate should be colorless and slightly alkahne. Add sulfuric acid till there is a slight excess. A colorless solution indicates the absence of acid magenta, while a light violet-red shows its presence, providing the amylalcohol extract showed no other dye to be present.

Caramel

Amthor's Test: 10 cc. of a solution of the fruit is put into a deep, narrow glass (a bottle may be used). Add 30 to 50 cc. of paraldehyde, to be gaged by the intensity of the coloring. Then add a sufficient quantity of absolute alcohol to make the solutions mix. If caramel is present, a brownish-yellow to dark-brown precipitate will be formed, decant, wash the precipitate once with absolute alcohol, dissolve in a little hot water and filter. The shade of color is proportional to the amount of caramel present. To verify the test, pour the colored fluid into a freshly prepared solution of phenylhydrazin (2 parts phenylhydrazin- hydrochloride, 3 parts sodium acetate, and 20 parts water). Much caramel produces a dark-brown precipitate in the cold, and is hastened by slightly heating. A very small amount of caramel will require several hours to precipitate.

Jams and Jellies

Apple-juice

Very often cider is added to other fruit juices to give them the proper consistency in jellies, jams, and marmalades. Its presence may sometimes be determined by making the usual starch test. A large quantity of starch is normally present in apples, but is less as they ripen, and finally disappears in the ripened fruit. There is no starch, or only a mere trace, in small fruits even when green. It is readily seen that if the juice is taken from green apples that there will be starch found in the artificial jelly or jam, though its absence does not prove the absence of cider.

Starch   In Jellies, Jams, and Such Products Make a solution of the jelly or jam and destroy the color by heating nearly to the boiling point and adding dilute (i : 3) sulfuric acid and potassium permanganate until the color is destroyed. This treatment does not affect the starch, and when cool add iodine, preferably potassium iodide-iodine (potassium iodide, 0.4 gram; iodin, o.i gram; water, 20 cc). If a great quantity of starch is present an almost black precipitate will be formed. Smaller amounts give the usual blue color. Whenever starch is found to be present, it is best to make a microscopical examination in the case of jams and marmalades. If the starch is normally present the grains will be seen within the cell walls after the iodine treatment. Starch is nearly always present in the apple and some other fruits, so unless it is present in jelly and such products in considerable quantity it is not likely that it was added.

Gelatin   **Henzold Test:** Add water to some of the jelly and boil for a short time, filter and treat the filtrate with an excess of a lo per cent solution of potassium bichromate and boil again. After cooling add 2 or 3 drops of concentrated sulfuric acid. A white flocculent precipitate forms if gelatin is present, and it gradually collects in a lump at the bottom.

E. Beckmann's Method: Treat the jelly with 95 per cent alcohol and wash the precipitate with alcohol to free it from the sugar, then drive off the alcohol by heating. Add a very little water to the residue and neutralize the extract with calcium carbonate. Then add formalin and evaporate to dryness. By this treatment gelatin is rendered insoluble. Pure fruit jellies have only 1 to 2 per cent of insoluble precipitate, while those jellies in which gelatin is used have 70 to 86 per cent of insoluble precipitate.

Agar agar Boil the sample with 5 per cent sulfuric acid. Add a crystal or two of potassium permanganate, and wait till it settles, and examine the sediment for diatoms with a microscope.

Their presence shows the use of agar.

Heavy metals in fruits and fruit products

Tin, Zinc, Lead, and Copper A. H. Allen's Method. — (See test for heavy metals under canned meat.)

Arsenic **Marsh's Test:** Fit a 100 cc. flask with a two-holed rubber stopper, through which passes a long-stemmed separatory funnel reaching nearly to the bottom, and a delivery tube which connects with a bulb tube containing a little acetate of lead solution. This in turn is connected with a calcium chloride tube and this with a small, hard glass tube, 15 or 20 cm. long, not over 0.5 cm. bore, and drawn to small size in the middle. The large part next the chloride tube is protected by fine wire gauze which extends to within a half inch of the constricted part. Two burners may be so placed as to heat the gauze. The flask should be placed in water and the bulb tube may be. Four grams of arsenic-free zinc, and 40 cc. of dilute pure sulfuric acid (i to 8) are placed in the flask. Let the hydrogen flow at least a quarter of an hour, then heat the gauze for 15 or 20 minutes. There should be no deposit in the tube. Now, char a portion of the sample, dissolve in water and pour into the separatory funnel, letting it run slowly into the flask. A dark deposit in the glass tube shoves that arsenic is present, but if after an hour no darkening takes place it is quite safe to say that no arsenic is present in the fruit.

Gutzei's Test: Place a gram of pure zinc, 5 cc. of dilute sulfuric acid (6 per cent) and about i cc. of a solution of the sample in a deep test tube. Cover the tube with three thicknesses of filter paper, fitted tightly over the mouth of the tube. Place on the upper paper a drop of strong silver nitrate solution. Place the tube in a dark place and leave for lo minutes. If a bright yellow stain forms on the filter paper, and turns black or brown when water is added to it, arsenic is present. Unless one is certain of the purity of the reagents used it is advisable to

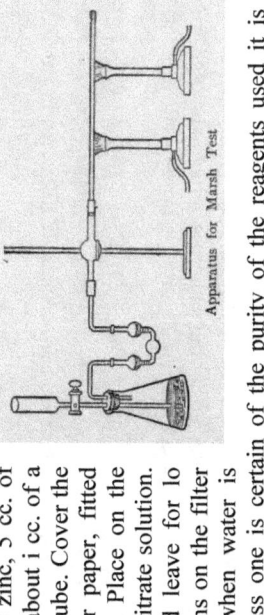

Apparatus for Marsh Test

make a blank test, using distilled water instead of the fruit. Oxidize all sulfids to sulfates before applying the above test. To find out whether they are present or not, substitute lead acetate for the silver nitrate on the filter paper. To avoid some of these difficulties treat according to the following method:

Preparation of the sample according to Leach: If possible reduce the sample to a dry char by treatment with concentrated nitric and sulfuric acids so that it may be powdered in a mortar. Dissolve out the arsenic by repeated treatment with boiling water. Save this extract, and, when cool, filter and submit to Marsh's test which is given above.

In case the sample is too much of a solid to get the arsenic out by the above treatment, it may be prepared according to the directions of Chittenden and Donaldson: Heat 100 grams of the macerated sample with 23 cc. of pure, strong nitric acid to a temperature of 150° C. or 160° C.

Assist the action by stirring occasionally. When the fruit becomes a deep yellow or orange color, remove the heat and add 3 cc. of pure, strong sulfuric acid. It should be stirred while nitrous fumes are passing off. Heat again to about 180° C, and before it cools add, drop by drop, 8 cc. of pure concentrated nitric acid. It should be stirred constantly while the acid is being added. Heat at 200° C. till sulfuric acid fumes begin to come off and only a dry mass remains. Powder the mass and exhaust it with boiling water, filter and test the solution when cold with Marsh's test.

Flavoring extracts (Lemon extract)

Lemon Oil
The presence of lemon oil may be detected by adding a large excess of water to a small amount of the extract in a test tube. If the mixture does not show some cloudiness, it is a strong indication that no lemon oil is present. The degree of cloudiness gives an idea of the amount of oil present.

Citral
This is present in the oil of lemon grass, which is sometimes used as a substitute for lemon oil. Citral may Be detected by the following test by Burgess: Add 20 cc. of sulfuric acid to 85 cc. of water. Dissolve in this mixture 10 grams of mercuric sulfate. Shake 2 cc. of the sample with 5 cc. of this reagent in a test tube. If citral is present, the liquid will be bright red, and will quickly disappear and give place to a whitish compound, which floats on top.

Oil of Citronella
This is often substituted for lemon oil. It may be detected by the same test which was used for citral. Instead of the red color and the white compound, citronella gives a bright yellow color, which does not disappear for some time.

Tartaric or Citric Acid
Precipitate the oil by the addition of an equal volume of water. Filter and add a very little of the filtrate to a test tube nearly full of cold lime water. A precipitate will form if tartaric acid is present and it will dissolve in an excess of ammonium chlorid or acetic acid. Filter, or, in case no

precipitate forms, heat the liquid. Citric acid is precipitated in the presence of a large excess of hot lime water.

Methyl Alcohol

Mullikin and Scudder. — Take 2 ft. of No. 12-15 copper wire and bend at right angles about 8 or 10 inches from one end. Grasp this bent end and an ordinary lead pencil side by side in such a way that the bend will be about the middle of the pencil. Wind the wire around the pencil and toward the free end of the short part of the wire until you have a coil 3 or 4 cm. long. Remove the pencil and twist the unwound parts together for a handle for the coil. Dilute a portion of the sample 3 or 4 times, and oxidize 10 cc. of the diluted liquid (preferably in a test tube) by heating to a red heat the above copper coil in the oxidizing flame of a Bunsen lamp. Thrust the heated coil quickly into the liquid contained in the test tube. In a second withdraw and immerse in water. Continue this operation till the oxide of copper fails to be reduced (usually 4 or 5 times is sufficient). Cool the liquid by immersing the tube in water. Separate into two parts and test each for formaldehyde by the following methods:

Mullikin, S. P test: Place one of these parts in an evaporating dish, and add to it 1 cc. of strong ammonia, boil gently over the free flame till the vapors cease to smell of ammonia. Add 2 or 3 drops of strong hydrochloric acid and heat just to boiling, and cool quickly by dipping the dish in cold water. Make the test for formaldehyde: Add a drop of a solution of resorcin (i: 200) and pour this mixture slowly down the side of an inclined test tube which contains 3 cc. of strong sulfuric acid, taking care to keep the liquids separate. After 3 minutes give the tube a rotary motion by rolling between the hands for a minute or more but only gradually mixing the water and acid, but mixing only about half of the acid. Flecks of a rose-red color form if methyl alcohol is present. Bands of color or flecks of other colors, even though they be tinged with red or a rose-red solution without the precipitate, should never be taken as proof that methyl alcohol is present. These conditions, however, are good grounds for repeating the test; 10 per cent or even less may be detected by this test.

Hydrochloric Acid and Ferric Chloride Test: Add a few drops of the other part of the above oxidized liquid to about 10 cc. of milk, known to be free from formaldehyde, in a porcelain casserole, and add 10 cc. of commercial hydrochloric acid (sp. gr. 1.2) which contains i cc. of 10 per cent ferric chlorid per 500 cc. Heat slowly over the open flame nearly to boiling. Give the liquid a rotary motion to break up the curd. If formaldehyde is present, the liquid will be colored violet. If not, it slowly turns brown. The presence of formaldehyde proves that methyl alcohol was in the original extract.

Flavoring extracts (Coloring matter): Preliminary Test: Treat the sample with strong hydrochloric acid, and if tropaeolin or methyl orange be present the solution will turn pink; Martins yellow partially decolorizes the solution; dinitrocresols decolorizes the solution. Turmeric or naphthol yellow produces no color change.

Turmeric   Turmeric may be detected by soaking a piece of filter paper in the sample, drying and dipping it in a dilute solution of boric acid or borax which has been slightly acidulated with hydrochlorid acid. Dry again and a cherry-red color forms if turmeric is present. Add a drop of dilute alkali and if turmeric be present the paper will be colored dark olive.

Coal-Tar Colors   Evaporate some of the extract to dryness; take up the residue with water and extract the coal-tar colors if present, and test for them by the method given under canned vegetables.

Flavoring extracts (Vanilla extracts): The best grades of vanilla extract are made by treating vanilla beans with 50 per cent alcohol. Coumarin, an extract from tonka beans, may be used in making the extract. This of course would make a cheaper product. If less than 50 per cent alcohol is used in making the extract, some alkali must be added to dissolve the resins which will not dissolve in a weaker alcohol. In artificial extracts some such coloring matter as caramel or tannin is used.

Preliminary Test: To a portion of the extract add a few drops of lead acetate solution. The absence of a bulky flocculent precipitate shows the extract not to be of high quality. Leach recommends that normal acetate of lead be added to the sample, and if a precipitate does not form it is conclusive evidence that it is not a pure extract. When a precipitate forms with the above reagent, it should settle immediately and leave a clear and almost colorless liquid. When there is a mere cloudiness only, it may be due to caramel, in which case the extract is to be suspected.

Alkali	Shake a portion of the sample with twice its volume of water. If no precipitate forms, an alkali is present. A flocculent reddish-brown precipitate shows no alkali is present. If the solution is milky it indicates the presence of a foreign resin. Add hydrochloric acid drop by drop to the diluted extract. Nothing more than a mere turbidity should result. Should it be quite turbid and the color fading after a time, it shows that an alkali has been used.
Foreign Resins	Mix a portion of the extract slowly with twice its volume of water, frequently shaking the mixture. When this solution is milky, it indicates a foreign resin.
	Hess Test: De-alcoholize 25 cc. of the sample by concentrating on the water-bath, adding water from time to time to retain the original volume. When no alkali is present in the extract, pure vanilla resin will be thrown down as a reddish-brown flocculent precipitate. Collect the resin, whatever its color, on a filter, and wash. Save the filtrate to test for caramel. Place a piece of the paper and resin in a dilute solution of potassium hydroxid. If the resin is that of pure vanilla it will dissolve, giving a deep red color, and is re-precipitated when the alkali is neutralized with

hydrochloric acid. Dissolve another part of the precipitate in alcohol, and to a part of this solution add a few drops of ferric chlorid; and to the other part, hydrochloric acid. There should be no marked coloration in either case if the resin is that of pure vanilla. Foreign resins nearly always produce a coloration.

Flavoring extracts (Caramel)

Tannin   Test another portion of the filtrate made in testing for foreign resins, with a few drops of a solution of gelatin. A slight precipitate only should form due to the presence of a small amount of tannin normally present in this filtrate. A large excess shows that it has been added to the extract.

Coumarin   **Leach's Test:** Dealcoholize a portion of the sample as above and treat with ammonia, add 3 or 4 volumes of chloroform in a separatory funnel. Evaporate the chloroform extract in an oven, not permitting the temperature to rise above 60° C. To the residue add a few drops of water; warm gently, and add a little of a solution of i gram of crystallized potassium iodide in 50 cc. of water, and the solution saturated with iodin. If coumarin is present, a brown, precipitate will form, and if stirred with a rod it will collect in dark green flecks.

Honey and saccharine products

Cane sugar   The presence of cane sugar can be detected with certainty only by the use of the polarimeter. Its presence in large quantity gives a high right-handed rotation.

Glucose syrup *Allen's Test* — Make the test for dextrin which is present in commercial glucose, but not in pure honey. Dilute a portion of the honey with an equal volume of water and add methyl alcohol with constant stirring until there is a permanent turbidity. If glucose syrup is present a heavy gummy precipitate will soon form. Genuine honey gives only slight milkiness.

Gelatin

Dilute a portion of the sample and add a solution of tannic acid. A precipitate indicates the presence of gelatin. Treat the sample with alcohol, and gelatin, if present, will be left undissolved, and it will give its characteristic odor on ignition.

Maple syrup

This is sometimes adulterated with glucose, molasses, golden syrup, and with ordinary white sugar. There are no satisfactory simple chemical tests for these substances. Pure maple syrup should have ash not lower than 0.35 to 0.40 per cent. A lower ash shows that cane sugar has been added. A higher ash would indicate the presence of molasses or brown sugar stock. These last two adulterants, if present in great abundance, may be detected by taste.

Glucose

This may be detected by the use of the polarimeter. Pure maple syrup gives 53.1 to 60 direct and 22.2 to 21.9 after hydrolysis. Maple syrup adulterated with glucose gives 80 to 100 direct and 18.9 to 45.6 after hydrolysis.

Spices

Flour

Boil 2 grams of the mustard in 4 or 5 cc. of distilled water for about 10 minutes. After it is cool, add a few drops of iodine solution slowly, avoiding a large excess though having a little uncombined iodine. If a blue color is produced, some starchy matter has been added to the mustard. The intensity of the reaction is an indication of the amount of starchy matter used. Pure mustard contains no starch and hence gives no reaction with iodine

Coloring matter

Pure mustard is a very light dull yellow, and whenever the sample is bright yellow, there is good grounds for suspecting the presence of some artificial coloring matter.

Turmeric

Add strong ammonium hydroxide to the mustard, and if turmeric is present an orange-red color is usually produced. Make an alcoholic extract of the sample and dip a piece of filter paper in it, and when dry draw it through a cold, saturated solution of boric acid in water. An orange or red-brown tint produced on the paper indicates the presence of turmeric. Thoroughly mix 2 or 3 grams of the mustard with castor oil and filter. If turmeric is present the filtrate will appear fluorescent. Extract a portion of the sample with 3 times its weight of wood alcohol and filter. Evaporate one half of the solution to dryness and add a little hydrochloric acid to the residue. This will turn red whenever turmeric is present, and if an excess of alkali be added it will change to a greenish blue. Evaporate the other half to dryness and moisten with a solution of boric acid and dry on a steam bath. A cherry-red color indicates turmeric.

Martius Yellow or Analogous Coal-Tar Coloring Matter

Extract the slightly acidified sample with 95 per cent alcohol and dye wool as directed under "Vegetables." The wool will be dyed a bright yellow.

Allen's Test: Treat a portion of the sample with cold alcohol, and shake vigorously for 5 minutes, then filter and evaporate the filtrate to dryness; add enough water to take up the residue and dye some white wool in this liquid as in the last test. When the dyed wool is wrapped in white paper and heated to 120° in an air bath, part of the coloring matter will be transferred to the paper. The coloring matter dissolves readily in dilute ammonia or hot water, and on the addition of hydrochloric acid the solution is decolorized and a yellow precipitate formed. This distinguishes it from picric acid.

Cayenne Pepper

Allen's Test: Boil a few gram of the mustard for a few minutes with alcohol, filter, and evaporate to dryness at about 100°. Taste the residue and cayenne may be recognized by its pungency. Or heat a portion of the extract, and smell the fumes. Irritation of the lungs and

Pepper

coughing will surely follow if cayenne pepper is present.

Pepper may be adulterated with wheat, buckwheat, pepper husks, ground olive stones, spent ginger. Cayenne pepper is sometimes added to adulterated pepper to give it the normal pungency. Many of these adulterants can be detected only by the aid of the microscope.

Neuss's Test: True pepper turns an intense yellow when covered with strong hydrochloric acid. Any adulteration can be detected at once by the color.

Ground Olive Stones or "Poivrette"

Make a paste of the pepper with caustic alkali. Dilute with a large quantity of water and wash by decantation. Olive stones will be colored a bright yellow; pepper-husks will appear dark.

Jumeau's Test: Dissolve 5 grams of iodin in a mixture of 50 cc. of ether and 50 cc. of alcohol. Cover the bottom of a porcelain capsule with the finely ground pepper, and add just enough of the iodin mixture to wet the entire mass, and mix well till it has the same consistency throughout. Let dry in the air, then powder and examine it, and if olive stones are present they will be colored yellow. Pure pepper would have a deep brown color. Aniline acetate, one part aniline in 3 parts acetic acid, colors pure pepper gray or white and olive stones yellowish brown.

Cayenne

Heat some of the red particles found in the pepper and their characteristic vapor is produced. Dissolve the particles in alcohol or ether and the same vapors are produced.

Vinegar

Vinegar

Vinegars may be adulterated by the addition of mineral acids as sulfuric or hydrochloric. Caramel or the coal-tar dyes may be employed to improve the color or to give color to an artificial product. Malic acid is always present in cider vinegar. Potassium acid tartrate occurs in true wine vinegar. Poisonous metals may be present in vinegars containing free mineral acid. Entirely artificial cider vinegar is often found on the market. If the vinegar is turbid from any suspended matter, it should be filtered. The samples should be analyzed at once, and in the laboratory they should always be kept in -glass-stoppered bottles.

General Observations: Ignite a little of the vinegar residue on a clean platinum wire in a colorless Bunsen flame, and if it is pure cider vinegar the flame will be colored the characteristic lilac color of potassium. The sodium flame is absent or only a mere trace of it is present. But in all artificially colored vinegars, spirit sugar and glucose vinegars, the sodium flame predominates. The residue of cider vinegar is thick, viscid, or mucilaginous, of a light brown color, astringent acid taste though not unpleasant. The solids of sugar-house vinegar, those from colored spirit and wood vinegar, each have a bitter taste on account of the caramel used to color them. The residue of the sugar-house vinegar has the odor of molasses. Wood vinegar when present gives a residue with a tarry or smoky taste and smell. Glucose vinegar gives the odor of scorched corn. Solids of fruit vinegars are quite soluble in alcohol, except a granular residue in grape vinegar, while the solids of malt and glucose vinegars are almost insoluble. The ash of fruit vinegars and malt vinegars has a distinct alkaline reaction, while that of spirit and wood vinegars is very feebly alkaline.

Free mineral acids

Ashby's Test: Extract 0.5 gram of logwood in 100 cc. of water and dry a drop or two on a porcelain surface. Then add a drop of the vinegar and dry again. If the residue is red, a mineral acid is present; if yellow, mineral acids are absent. When only a very small amount of the acid is present the red coloration will be destroyed on diluting with water, but may be restored by concentrating the liquid.

Sulfuric Acid	Sulfuric acid, if present, will cause the vinegar to leave a charred mass when evaporated over the water-bath.
Vinegar	**Frear's Method:** Mix 5 cc. of the sample and 5 or 10 cc. of water, and add a very little of a solution of methyl violet (made by dissolving one part of methyl violet 2 B. in 100,000 parts of water). A blue or green coloration shows the presence of mineral acids. Sulfuric Acid as Distinguished from Sulfates Attends Method. — Evaporate 100 cc. of the vinegar down to one tenth its volume, and when cold add 50 cc. of alcohol. Sulfuric acid remains in solution while the sulfates are precipitated. Dilute the solution and precipitate the acid with barium chloride.
Hydrochloric acid free	Place a definite quantity of the vinegar in a distilling flask and distil ofif half. Add a few drops of silver nitrate to the distillate. If a precipitate forms, hydrochloric acid is present.
Malic acid	Leaches Method. — To 5 cc. of the sample, add a few drops of a solution of calcium chlorid (i: 10); make slightly alkaline with ammonia. Filter off any precipitate that may form, add 20 to 30 cc. of 95 per cent alcohol to the filtrate and heat to boiling. If maUc acid is present, a voluminous flocculent precipitate will form. A precipitate may form in vinegars containing dextrine. Make a further test for malic acid by the following: Filter and treat the precipitate with a little alcohol, and when dry add concentrated nitric acid and evaporate to dryness on a waterbath. Treat the residue with sodium carbonate, boil for a short time, filter. Add acetic acid to the filtrate till slightlyalkaline, boil till carbon dioxid is expelled, and if on the addition of calcium sulfate a precipitate forms, it indicates the presence of malic acid.

Coloring matter

Caramel

Crampton and Simons method: Shake well togetherin a corked flask 50 cc. of the vinegar with about half as many grams of fullers' earth; after standing for half an hour filter. Vinegar containing no artificial color will show scarcely any change in color when thus treated. A caramel colored vinegar will be decolorized in proportion to the amount of caramel present.

Coal-Tar Colors in Wine Vinegar

Test by the usual test for coal-tar dyes. See under canned vegetables.

Metallic impurities

Vinegars containing free mineral acids are sometimes found to contain poisonous metals. Evaporate 200 to 400 cc. of the vinegar to dryness, add a little sodium hydroxide to this residue and burn to an ash over a low flame. It may be necessary to add a little potassium nitrate once or twice. Add a little dilute hydrochloric acid and saturate with hydrogen sulfide and test for lead, zinc, copper, and arsenic according to Allen's method given under canned meats.

Vinegar (Spices to increase pungency)

Leach: Neutralize a portion of the vinegar with sodium carbonate. The presence of spices is easily detected by tasting this mixture.

Another Test: Exactly neutralize a little of the vinegar as above, evaporate to smaller bulk and taste as before, then shake the concentrated liquid with ether, separate the ethereal layer and evaporate it, and taste the residue.

Tartar in Wine Vinegar

Allen's Method: Evaporate a portion of the vinegar and treat the residue with alcohol; a granular residue of tartar remains undissolved. To prove that it is tartar, decant the alcohol and dissolve the residue in a little hot water, cool, rub the inside of the vessel with a glass rod, and if tartar was present acid potassium tartrate will be deposited where the rod touched the vessel. The test will be more sensitive if an equal volume of alcohol is added.

Free Tartaric Acid in Wine Vinegar	Test as for Tartar. — Treat the alcoholic solution of the extract with an alcoholic solution of potassium acetate. Rub the sides of the vessel as before, and if tartaric acid is present the streaks and sometimes a precipitate forms where the rod touches the vessel.
Glucose	Glucose is present when both direct and invert readings are dextrorotatory.

Fats and Oils

Cottonseed oil and Cottonseed stearin	Lard is very often adulterated with cottonseed oil, cottonseed stearin and beef stearin. Their being very much cheaper accounts for the sophistication.
	Halphen's Test — Dissolve 1 per cent of sulfur in a given volume of carbon bi sulfide. Add an equal volume of amylic alcohol. Mix 3 to 5 cc. of this reagent with an equal volume of the melted lard in a test tube. Close with a cotton stopper and boil for 15 minutes in a bath of saturated brine. The presence of cottonseed oil is indicated by a deep-red or orange color, little or no color resulting in its absence. Lard from hogs fed on any of the various cottonseed products may give a faint reaction when this test is applied.
Beef stearin	It is very difficult to identify beef-stearin by chemical tests. It is usually detected by use of the microscope. Make a solution of 2 to 5 grams of the fat in 10 to 20 cc. of ether. Let stand a half day, at about the room temperature. Loosely stopper the tube with cotton to prevent too rapid evaporation of the ether.
	It is well to vary the conditions of heat, amount of solvent, and rate of crystallization, to get the best possible results. It may often be well to separate the crystals thus obtained by filtering and

re-crystallizing from ether. Separate the crystals that form at the bottom of the test tube from the liquid portion by pouring on a small filter. Wash them several times with ether, but not sufficient to remove the mother liquor entirely. In case it is all removed, and the crystals are too fragile to mount, add a drop of alcohol.

Crystals of lard stearin are flat rhomboidal plates, one end being oblique to the sides, and they do not appear to be regularly grouped. Beef-stearin crystals are rod-shaped, or needles often apparently curved with pointed ends, and are arranged in clusters like the ribs of a fan, the crystals radiating from a common point. Under certain conditions the lard crystals are not irregularly grouped, but are arranged like the parts of a feather, where one part seems attached to another close at hand. Considerable experience is necessary to use this test with absolute certainty.

Olive Oil

Olive oil is one of the most commonly adulterated foods. The commonest adulterant probably is cottonseed oil. Other foreign oils, such as peanut, sesame, and rape, are sometimes used.

Preliminary Test— Pure olive oil turns from a pale to a dark-green color in a few minutes, when it is shaken with the same volume of concentrated nitric acid or sulfuric acid. Whenever reddish to orange or brown coloration results, the presence of a foreign vegetable oil is indicated (probably seed oil). Olive oil when shaken with nitric acid gives a pale reen, which changes to an orange yellow after heating five minutes. With similar treatment peanut oil gives pale rose and brownish yellow; rape oil, pale rose and orange yellow; sesame oil, white and brownish yellow; sunflower oil, dirty white and reddish yellow; cottonseed oil, yellowish brown and reddish brown; castor oil, pale rose and golden yellow.

Pontevs Test; Elaiden Test — Treat 1 cc. of mercury with 12 cc. of cold nitric acid (sp. gr. 1.42) and shake 2 cc. of this freshly-made solution with 50 cc. of the sample in a bottle every 10 minutes for 2 hours. Oils which are principally olein, or mixtures of olein and solid esters like palmatin and stearin, give more or less solid products, but olive oil is remarkable for the firmness of the canary or lemon-yellow mass which is formed. After standing a day, the mass cannot be pierced with a glass rod and sometimes it gives forth a sound when struck. This test requires considerable experience to be used with any great degree of certainty.

Cottonseed oil

Carbon bi sulfide containing 1 per cent of sulfur in solution is mixed with an equal amount of amyl alcohol. Equal volumes (about 3 cc. of this reagent and the sample, are mixed in a test tube and heated in a bath of boiling saturated brine for a quarter of an hour. The presence of cottonseed oil is shown by the formation of a deep-red or orange color. Little if any color is produced in its absence. If no color is produced it is well to add another cc. of the reagent and heat 5 or 10 minutes more, and to repeat this again if no color forms. Lard and lard oil from animals fed on cottonseed meal may give a faint reaction.

Peanut Oil

Bellier's Test: Saponify a gram of the sample with 5 cc. of a solution of 85 grams potassium hydroxide in a liter of strong alcohol. This may be done in a small Erlenmeyer flask on the water-bath. Then boil for two minutes, neutralize exactly with dilute acetic acid (use phenolphthalein as the indicator). Cool the mixture by placing the flask in water at 17° to 19° C. A precipitate usually forms. Add 50 cc. of 70 per cent alcohol which contains one per cent by volume of concentrated hydrochloric acid (sp. gr. 1.2). Shake the flask vigorously and cool again as before. If no precipitate forms the oil is not adulterated with peanut oil. The presence of 10 per cent or more of peanut oil produces a precipitate; even a smaller amount will produce cloudiness after standing between 17° and 19° C. for 30 minutes. To distinguish these from peanut oil heat the mixture on the water-bath till everything has dissolved, and cool to 17° to 19°. The cloudiness will not appear if the oil is pure, but will reappear if peanut oil is present.

Sesame oil

Badouin's Test: About 0.1 of a gram of cane sugar is dissolved in 10 cc. of hydrochloric acid (sp. gr. 1.20) shaken vigorously with 20 grams of the sample for a minute or more. After standing for a while the aqueous solution will separate from the oil. If 1 per cent or more of sesame oil is present, the aqueous solution will be colored crimson.

Tocher's Test: Dissolve 1 gram of pyrogallic acid in 15 cc. of strong hydrochloric acid. Add an equal volume of the oil in a separating funnel. When it has stood a minute, draw off the aqueous solution and boil. In the presence of sesame oil it is colored red by transmitted light, and blue by reflected light.

Rape Oil

Palas' Test — Make a 1 per cent solution of fuchsin and a 30 per cent solution of sodium acid sulfite. Mix together 20 cc. of each of these solutions and add 200 cc. of water and 5 cc. of strong sulfuric acid. After the solution is decolorized, 10 cc. of the sample is shaken with it. If rape oil is present, the color will be partially restored. To prevent the formation of the color by contact with the air have the vessel full of the mixture.

Beverages

Coffee

Coffee is often colored with such substances as Scheele's green, chrome yellow, iron oxide, Prussian blue, indigo and turmeric. Imitation coffee beans have been made of wheat flour, bran, rye, chicory and peas.

Allen's Preliminary Test: A good preliminary test for ground coffee is to sprinkle some of it on the surface of cold water. The oil of true coffee prevents the particles from being readily soaked, and so they float for some time. Chicory and most of the other adulterants of coffee contain no oil, but do contain caramel, which is quickly extracted by the water producing a zone of brown color about such particles. They become soaked and quickly sink. The liquid containing pure coffee diffuses uniformly without coloring the water to any perceptible degree. Chicory and similar roots give a dark brown, turbid infusion. Roasted cereals do not impart so distinct a color to water.

Coffee - Colouring Matter

Shake the coffee beans in cold water and make the regular qualitative tests for the inorganic coloring matters — Scheele's green may be identified by testing for copper and arsenic; chrome yellow, by testing for lead chromate; iron oxide may be detected by its characteristic tests. Organic coloring matter is best extracted with alcohol. Prussian blue may be detected by dissolving it from the sediment with hot caustic alkali, acidifying with hydrochloric acid, treating it with a drop of ferric chloride. If present, ferric ferrocyanide, a blue precipitate, will be formed. Indigo is not discharged by sodium hydroxide, while Prussian- blue is. It will form a deep blue solution with sulfuric acid.

Imitation Coffee Beans

Most imitation coffee, as already stated, is heavier than water. Coffee contains no starch, so the imitation beans made of cereals may be detected by testing for starch.

Coffee - Starch

Allen's Method: Boil the coffee in lo parts of water. When perfectly cold add to it a little sulfuric acid, then a strong solution of potassium permanganate, drop by drop, with constant shaking, till the liquid is almost decolorized; strain or decant and add to the solution a solution of iodine. If 1 per cent or more of starch is present, a blue coloration will be produced.

Coffee - Chicory

Rimmington's Test: Boil a portion of the sample with water which contains a little sodium carbonate; decant, wash and treat the residue with a weak solution of bleaching powder for several hours. The solution will be decolorized. The coffee will be at the bottom as a dark layer while the chicory will be a light layer above it.

Albert Smith's Test: Boil lo grams of the sample in 250 cc. of water; strain and add basic lead acetate in slight excess. A precipitate forms, and when it has settled the supernatant liquid will be colorless if the coffee is pure, but more or less colored if chicory is present.

Tea - Foreign leaves

Though there are several chemical tests for foreign leaves, none are as satisfactory as a microscopic examination. Soften the leaves by soaking in hot water, unroll carefully and examine with a hand lens or low power of the microscope. Compare with a genuine leaf — the shape, margin, and venation.

Exhausted tea leaves

Sometimes such leaves may be detected by a physical examination. They are often more or less unrolled and broken on the edges. But the only certain way of ascertaining their presence is to determine the soluble ash which is from 2.5 to 4 per cent in pure tea and usually less than 0.8 per cent in exhausted tea.

Tea - Facing

Shake the tea leaves in cold water and make the regular qualitative tests for the inorganic coloring matters. Scheele's green may be identified by testing for copper and arsenic; chrome yellow, by testing for lead chromate; iron oxide may be detected by its characteristic tests. Organic coloring matter is best extracted with alcohol. Prussian blue may be detected by dissolving it from the sediment with hot caustic alkali, acidifying with hydrochloric acid, treating it with a drop of ferric chloride. If present, ferric ferrocyanide, a blue precipitate, will be formed.

Indigo is not discharged by sodium hydroxide, while Prussian-blue is. It will form a deep blue solution with sulfuric acid.

Tea - Catechu *Hager's Test:* Boil a little of the tea in water, and add to the extract an excess of lead monoxide. If the tea is pure the addition of a solution of silver nitrate produces only a slight grayish precipitate, but when catechu is present a yellow flocculent precipitate forms.

Simple Screening test for Detecting Adulteration in Common Food

(Source: http://agmarknet.nic.in/adulterants.htm)

Food article	Adulteration	Test
Vegetable oil	Castor oil	Take 1 ml. of oil in a clean dry test tube. Add 10 ml. Of acidified petroleum ether. Shake vigorously for 2 minutes. Add 1 drop of Ammonium Molybdate reagent. The formation of turbidity indicates presence of Castor oil in the sample.
	Argemone oil	Add 5 ml, conc. HNO_3 to 5 ml. sample. Shake carefully. Allow to separate yellow, orange yellow, crimson colour in the lower acid layer indicates adulteration.
Ghee	Mashed Potato Sweet Potato, etc.	Boil 5 ml. Of the sample in a test tube. Cool and a drop of iodine solution. Blue colour indicates presence of Starch. colour disappears on boiling & reappears on cooling.
	Vanaspati	Take 5 ml. Of the sample in a test tube. Add 5 ml. Of Hydrochloric acid and 0.4

	ml of 2% furfural solution or sugar crystals. Insert the glass stopper and shake for 2 minutes. Development of a pink or red colour indicates presence of Vanaspati in Ghee.
Rancid stuff (old ghee)	Take one teaspoon of melted sample and 5 ml. Of HCI in a stoppered glass tube. Shake vigorously for 30 seconds. Add 5 ml. Of 0.1% of ether solution of Phloroglucinol. Restopper & shake for 30 seconds and allow to stand for 10 minutes. A pink or red colour in the lower(acid layer) indicates rancidity.
Synthetic Colouring Matter	Pour 2 gms. Of filtered fat dissolved in ether. Divide into 2 portions. Add 1 ml. Of HCI to one tube. Add 1 ml. Of 10% NaOH to the other tube. Shake well and allow to stand. Presence of pink colour in acidic solution or yellow colour in alkaline solution indicates added colouring matter.
Honey	Fiehe's Test: Add 5 ml. Of solvent ether to 5 ml. Of honey. Shake well and decant the ether layer in a petri dish. Evaporate completely by blowing the ether layer. Add 2 to 3 ml. Of resorcinol (1 gm. Of resorcinol resublimed in 5 ml. Of conc. HCI.) Appearance of cherry red colour indicates presence of sugar/jaggery.
Invert sugar/jaggery	Aniline Chloride Test : Take 5 ml. Of honey in a porcelain dish. Add Aniline Chloride solution (3 ml of Aniline and 7 ml. Of 1:3 HCI) and stir well. Orange red colour indicates presence of sugar.
Pulses/Besan	
Kesari dal(Lathyrus sativus)	Add 50 ml. Of dil.HCI to a small quantity of dal and keep on simmering water for about 15 minutes. The pink colour, if developed indicates the presence of Kesari dal.

Pulses	Metanil Yellow(dye)	Add conc. HCl to a small quantity of dal in a little amount of water. Immediate development of pink colour indicates the presence of metanil yellow and similar colour dyes.
	Lead Chromate	Shake 5 gm. Of pulse with 5 ml. Of water and add a few drops of HCl. Pink colour indicates Lead Chromate.
Bajra	Ergot infested Bajra	Swollen and black Ergot infested grains will turn light in weight and will float also in water
Wheat flour	Excessive sand & dirt	Shake a little quantity of sample with about 10 ml. Of Carbon tetra chloride and allow to stand. Grit and sandy matter will collect at the bottom.
	Excessive bran	Sprinkle on water surface. Bran will float on the surface.
	Chalk powder	Shake sample with dil.HCl Effervescence indicates chalk.
Common spices like Turmeric, chilly, curry powder,etc.	Colour	Extract the sample with Petroleum ether and add 13N H_2SO_4 to the extract. Appearance of red colour (which persists even upon adding little distilled water) indicates the presence of added colours. However, if the colour disappears upon adding distilled water the sample is not adulterated.
Black Pepper	Papaya seeds/light berries, etc.	Pour the seeds in a beaker containing Carbon tetra-chloride. Black papaya seeds float on the top while the pure black pepper seeds settle down.
Spices	Powdered bran	Sprinkle on water surface. Powdered bran and sawdust float on the surface.

(Ground) and saw dust		
Coriander powder	Dung powder	Soak in water. Dung will float and can be easily detected by its foul smell.
	Common salt	To 5 ml. Of sample add a few drops of silver nitrate. White precipitate indicates adulteration.
Chillies	Brick powder grit, sand, dirt, filth, etc.	Pour the sample in a beaker containing a mixture of chloroform and carbon tetrachloride. Brick powder and grit will settle at the bottom.
Badi Elaichi seeds	Choti Elaichi seeds	Separate out the seeds by physical examination. The seeds of Badi Elaichi have nearly plain surface without wrinkles or streaks while seeds of cardamom have pitted or wrinkled ends.
Turmeric Powder	Starch of maize, wheat, tapioca, rice	A microscopic study reveals that only pure turmeric is yellow coloured, big in size and has an angular structure. While foreign/added starches are colourless and small in size as compared to pure turmeric starch.
Turmeric	Lead Chromate	Ash the sample. Dissolve it in 1:7 Sulphuric acid (H_2SO_4) and filter. Add 1 or 2 drops of 0.1% dipenylcarbazide. A pink colour indicates presence of Lead Chromate.
	Metanil Yellow	Add few drops of conc.Hydrochloric acid (HCl) to sample. Instant appearance of violet colour, which disappears on dilution with water, indicates pure turmeric. If colour persists Metanil yellow is present.

IAPEN | *Alphabetic list of food adulterants*

Cumin seeds (Black jeera)	Grass seeds coloured with charcoal dust	Rub the cumin seeds on palms. If palms turn black adulteration in indicated.
Asafoetida (Heeng)	Soap stone, other earthy matter	Shake a little quantity of powdered sample with water. Soap stone or other earthy matter will settle at the bottom.
	Chalk	Shake sample with Carbon tetrachloride (CCl_4). Asafoetida will settle down. Decant the top layer and add dil.HCl to the residue. Effervescence shows presence of chalk.
Food grains	Hidden insect infestation	Take a filter paper impregnated with Ninhydrin (1% in alcohol.) Put some grains on it and then fold the filter paper and crush the grains with hammer. Spots of bluish purple colour indicate presence of hidden insects infestation

SIMPLE SCREENING TEST FOR DETECTING ADULTERATION IN COMMON FOOD

(Quick Test for some adulterants in foods, Food Safety and Standards Authority of India, Government of India)
Download Link: http://www.fssai.gov.in/Portals/0/Pdf/Final_test_manual_part_1%2816-08-2012%29.pdf

Food article	Adulteration	Test
Milk	Water	i. The lactometer reading shall not ordinarily be less than 26. ii. The presence of water can be by putting a drop of milk on a polished slanting surface. The drop of pure milk either or flows slowly leaving a white trail behind it, whereas milk adulterated water will flow immediately without leaving a mark.
	Starch	Add a few drops of tincture of Iodine or Iodine solution. Formation of blue colour indicates the presence of starch.
	Removal of Fat	The lactometer reading will go above 26 while the milk apparently remains thick.
Milk	Glucose	Take a teaspoonful of milk in a test tube. Dip a strip of diastix in it or 30 second. A change in colour from blue to green indicates the presence of glucose in milk.
	Sugar	Take 3 ml of the milk in a test tube. Add 2 ml of the hydrochloric acid. Heat the

test tube after adding 50 mg of resorcinol. The red colouration indicates the use of sugar in the milk.

Sodium -bi-carbonate/ Neutralizer	Take 3 ml of the milk in a test tube and add 5 mL of rectified spirit to it. Then add 4 drop of rosalic acid solution. The appearance of red/rosy colouration indicates the presence of sodium- bi- carbonate in the milk.
Urea	Take a teaspoon of milk in a test tube. Add ½ teaspoon of soybean or arhar powder. Mix up the contents thoroughly by shaking the test tube. After 5 minutes, dip a red litmus paper in it. Remove the paper after½ a minute. A change in colour from red toblue indicates the presence of urea in the milk.
Boric acid	Take 3 ml of milk in a test tube. Add 20 drops of hydrochloric acid and shake the test tube or mix up the contents thoroughly. Dip a yellow paper-strip, and remove the same after 1 minute. A change in the colour from yellow to red, followed by the change from the red to green , by addition of ammonia drop solution, indicates that the boric acid is present in the milk (to prepare the yellow paper-strip, dips strips of filter paper in an aqueous solution of the turmeric , and dry it up).
Vanaspati	Take 3 ml of milk in a test tube. Add 10 drops of hydrochloric acid. Mix up one teaspoonful of sugar. After 5 minutes, examine the mixture. The red colouration indicates the presence of vanaspati in the milk.

	Formalin	Take 10 ml of milk in a tests tube and add 5ml of con sulphuric acid from the sides of the wall without shaking. If a violet or bluering appears at the intersection of two layers then it shows presence of formalin.
	Detergent	Shake 5-10 ml. of sample with an equal amount of water lather indicates the presence of detergent.
	Sodium chloride	Take 2 ml of milk in a test tube. Add 0.1 ml of 5% potassium chromate solution and 2.0 ml of 0.1 N silver nitrate. Appearance of red precipitate indicates the presence of sodium chloride in milk.
Milk	Synthetic milk	Synthetic milk has a bitter after taste, gives a soapy feeling on rubbing between the fingers and turns yellowish on heating.
	Synthetic milk-test for protein	The milk can easily be tested by Urease strips (available in the Medical stores) because Synthetic milk is devoid of protein.
	Test for glucose /inverted sugar/sugar syrup.	Milk does not contain glucose /invert sugar, if test for glucose with urease strip found positive. It means milk is adulterated.
	Sodium Chloride	Take 2 ml of milk in a test tube, ad2drops of 5% Potassium chromate solution and 2 ml of 0.1 N silver nitrate solutions to it. The appearance of a red precipitate indicates the absence of dissolved chloride in milk and appearance of yellow colour indicates the presence of dissolved chloride.

Sweet Curd	Vanaspati	Take1 teaspoon full of curd in a test tube. Add 10 drops of hydrochloric acid. Mix up the contents shaking the test tube gently. After 5 minutes, examine the mixture. The red colouration indicates the presence of vanaspati in the curd.
Rabdi	Blotting paper	Take a teaspoon of rabri in a test tube. Add 3 ml of hydrochloric acid and 3 ml of distilled water. Stir the content with a glass rod. Remove the rod and examine. Presence of fine fibres to the glass rod will indicate the presence of blotting paper in rabri.
Khoa and its products	Starch	Boil a small quantity of sample with some water, cool and add a few drops of Iodine solution. Formation of blue colour indicates the presence of starch.
Chhana or Paneer	Starch	Boil a small quantity of sample with some water, cool and add a few drops of Iodine solution. Formation of blue colour indicates the presence of starch.
Ghee, cottage cheese, condensed milk, khoa, milk powder etc,	Coal Tar Dyes	Add 5 ml of dil. H_2SO_4 or conc. HCL to one teaspoon full of melted sample in a test tube. Shake well. Pink colour (in case of H_2SO_4) or crimson colour (in case of HCl) indicates coal tar dyes. If HCl does not give colour dilute it with water to get the colour.
Ghee	Vanaspathy or Margarine	Take about one tea spoon full of melted sample of ghee with equal quantity of concentrated Hydrochloric acid in a stoppered test tube and add to it a pinch of sugar. Shake for one minute and let it for five minutes. Appearance of

crimson colour in lower (acid) of Vanaspati or Margarine.

	Mashed Potatoes, Sweet Potatoes and other starches	The presence of mashed potatoes and sweet potatoes in a sample of ghee can easily be detected by adding a few drops of Iodine, which is brownish in colour turns to blue if mashed potatoes/sweet potatoes/other starches are present.
Butter	Vanaspati or Margarine	Take about one teaspoon full of melted sample of butter with equal quantity of concentrated Hydrochloric acid in a stoppered test tube and add to it a pinch of sugar. Shake for one minute and let it for five minutes. Appearance of crimson colour in lower (acid)layer shows presence of Vanaspati or Margarine.
	Mashed potatoes other starches	The presence of mashed potatoes and sweet potatoes in a sample of butter can easily e detected by adding a few drops of iodine (which is brownish in colour), turns to blue.
Oils and fats	Argemone oil	Take small quantity of oil in a test tube. Add equal quantity of concentrated Nitric acid and shake carefully. Red to reddish brown colour in lower (Acid) layer would indicate the presence of Argemone oil.
	Mineral oil	Take 2 ml of the oil sample and add an equal quantity of N/2 Alcoholic potash. Heat in boiling water bath (dip in boiling water) for about 15 minutes and add 10 ml of water. Any turbidity shows presence of mineral oil.
	Castor oil	Take about one ml of the oil, add 10 ml of acidified petroleum ether and mix well, Add a few drops of ammonium molybdate reagent. Immediate appearance

of white turbidity indicates the presence of castor oil.

TOCO (Tri-ortho-cresyl-phosphate)	Take 2 ml of suspected sample of oil. Add a little butter yellow crystal. Immediate formation of red colour indicates presence of TOCP.
Mustard oil	Take about 3 ml of mustard oil in a test tube. Add 2 ml of amyl alcohol in it and 1 ml of carbon di -sulphide and a little amount of sulphur. Plug the mouth of the test tube and heat it on the flame of a spirit lamp for 3 minutes. A red colouration indicates the presence of cotton seed oil in the mustard oil.
Cotton seed oil	Take 20 drops of the edible oil in each of the four test tubes. Make 3 different solutions, mixing up 1 part of distilled water, 3 parts of distilled water and 4 parts of distilled water. Add 2 ml of each solution in each of the test tubes and add 2 ml of hydrochloric acid in the mixture of any tube, indicates the presence of prohibited colour in the edible oil.
Edible oil Prohibited colour	
Rancidity	Take 3 ml of the edible oil in a test tube.Add 3 ml of hydrochloric acid, in it. Close the mouth of the test tube. Mix up the content. Add 3ml of 0.1% Phloroglucinol solution in it. Shake the test tube vigorously for 2 minutes and keep it aside. Examine the test tube after 30 minutes. A pink or red colouration in acid layer indicates that, the oil sample is rancid.
Karanja oil (Pungam oil).	To two drops of the oil add a solution of 40% antimony trichloride in chloroform. Appearance of yellow to orange colour immediately shows the presence of Karanja oil.

Coconut oil	Any other oil	Place a small bottle of oil in refrigerator. Coconut oil solidifies leaving the adulterant as a Separate layer.
	cyanide	Take 3ml of the edible oil in a test tube. Add 10 drops of alcoholic potash and heat the tube on the flame of a spirit lamp. Make an addition of a little amount of each of the ferrous sulphate and ferric chloride in the test tube, and shake it to mix up the contents thoroughly. Add 3 ml hydrochloric acid. The blue colouration indicates the presence of hydro cyanide acid, which gets produced due to presence of cyanide in edible oil.
Sugar	Chalk powder	Dissolve 10 gm of sample in a glass of water, allow settling, Chalk will settle down at the bottom.
	Urea	On dissolving in water it gives a smell of ammonia.
Pithi Sugar	Washing Soda	Add few drops of Hydrochloric acid, effervescence (give off bubbles) will indicate the presence of washing soda.
	Chalk powder	Dissolve 10 gm of sample in a glass of water, allow to settle, chalk will settle down at the bottom.
	Yellow colour (Non- permitted)	Take 5 ml in a tests tube from the above solution and add a few drops of conc. Hcl. A pink colour in lower acid layers shows the presence of non-permitted colour.
Honey	Sugar solution	A cotton wick dipped in pure honey when lighted with a match stick burns and shows the purity of honey. If adulterated, the presence of water will not

Jaggery	Sodium bicarbonate	allow the honey to burn, If it does; it will produce a cracking sound. Take ¼ of a tea spoon of the jaggery in a test tube. Add 3 ml of Muratic acid. The presence of sodium carbonate effects effervescence.
	Metanil yellow colour	Take ¼ of a teaspoon of the jaggery in a test tube. Add 3 ml of alcohol and shake the tube vigorously to mix up the content. Pour 10 drops of hydrochloric acid in it. A pink colouration indicates the presence of metanil yellow colours in jaggery.
Jaggery	Washing soda	Add a few drops of solution HCl. Effervescence shows presence of washing soda.
	Chalk powder	Dissolve a little amount sample in water in a test tube, chalk powder settles down or add a few drops of conc HCl solution, effervescence indicates the presence of adulterant.
Honey	Invert sugar/Jaggery	Fiehe's Test: Add 5 ml. of solvent ether to 5 ml. of honey. Shake well and decant the ether layer in a Petri dish. Evaporate completely by blowing the ether layer. Add 2 to 3 ml. of resorcinol (1 gm. Of resorcinol resublimed in 5 ml. Of conc. HCl.) Appearance of cherry red colour indicates presence of sugar/jaggery.
	Sugar solution	Add a drop of honey to a glass if water, if the drop does not disper se in water it indicates that the honey is pure. However, if the drop disperses in water it indicates presence of added sugar.

Bura sugar	Washing soda	Add 1 ml of HCl to a little of bura sugar. Effervescence occurs if washing soda is present. Dissolve 2 gm of sugar in water; dip a red litmus paper in the solution. If washing soda is present, it will turn blue.
Sweetmeats, Ice-cream and beverages	Metanil yellow (a non-permitted coal tar colour)	Extract colour with luke-warm from food articles. Add few drops of concentrated Hydrochloric acid. If magenta red colour develops the presence of metanil yellow is indicated.
	Saccharin	Taste a small quantity. Saccharin leaves a lingering sweetness on tongue for a considerable time and leaves a bitter taste at the end. Take two spoons of liquid sample or about 5 to 10 gms of solid sample with little quantity of water in a test tube, add few drops of Hydrochloric acid and 10 ml of solvent ether. Shake well. Decant the ether layer into a test tube or a beaker, evaporate the ether spontaneously. Add one drop of water (warm) to the residue and taste. Sweet taste will indicate the presence of saccharin
Wheat, Rice, Maize, Jawar, Bajra, channa, Barley etc.	Dust, pebble, stone, straw, weed seeds, damaged grain, weevilled grain, insects, rodent hair and excreta	These may be examined visually to see foreign matter, damaged grains, discoloured grains, insect, rodent contamination etc.

Ergot (a fungus containing poisonous substance)	(i) Purple black longer sized grains in Bajra show the presence of Ergots. (ii) Put some grains In a glass tumbler containing 20 per cent salt solution Ergot floats over the surface while sound grains settle down.
Dhatura	Dhatura seeds are flat with edges with blackish brown colour which can be separated out by close examination.
Karnel Bunt	The affected wheat kernel have a dull appearance, blackish in colour and rotten fish smell,
Sella Rice (Parboiled Rice) Metanil yellow (a non-permitted coal tar colour)	Rub a few grains in the palms of two hands. Yellow would get reduced or disappear. Add a few drops of dilute Hydrochloric acid to a few rice grains mixed with little water, presence of pink colour indicates presence of Metanil yellow
Turmeric (colouring for golden appearance)	(i) Take a small amount of sample in a test tube, add some water and shake. Dip Boric acid paper (filter paper dipped in Boric acid solution) If it turns pink turmeric is present. (ii) Take some rice and sprinkle on it a small amount of soaked lime for some time, grains will turn red if turmeric is present.
Parched rice	Urea	Take 30 numbers of parched rice in a test tube. Add 5ml of distilled water in it.

		Mix up the contents thoroughly, by shaking the test tube. After 5 minutes, filter the water- contents, and add ½ teaspoon of powder of arhar or soybean in it. Leave it for 5 minutes, and then dip a red litmus paper in the mixture. Take out the litmus paper after 30 seconds and examine it. A blue colouration indicates the presence of urea in the parched rice.
Maida/ Rice	Boric Acid	Take a small amount of sample in a test tube, add some water and shake. Add a few drops of HCl. Dip a turmeric paper strip if it turns red, boric acid is present.
Maida	Resultant atta or cheap flour	When dough is prepared from resultant or left out atta, more water has to be used. The normal taste of chapattis prepared out of wheat is somewhat sweetish whereas those prepared out of adulterated wheat will taste insipid.
Food grains	Hidden insect infestation	Take a filter paper impregnated with Ninhydrin (1% in alcohol.) Put some grains on it and then fold the filter paper and crush the grains with hammer. Spots of bluish purple colour indicate presence of hidden insects infestation
Wheat flour	Excess bran	Sprinkle on water surface. Bran will float on the surface.
Wheat flour	Chalk powder	Shake sample with dil. HCl Effervescence indicates chalk.
	Excessive sand and dirt	Shake a little quantity of sample with about 10 ml. Of Carbon tetra chloride and allow to stand. Grit and sandy matter will collect at the bottom.
Dal whole and spilt	Khesari Dal	(i) Khesari dal has edged type appearance showing a slant on one side and square in appearance in contrast to other dals.

Food	Adulterant	Detection
		(ii) Add 50 ml of dilute Hydrochloric acid to the sample and keep on simmering water for about 15 minutes. The pink colour developed indicates the presence of Khesari dal.
	Clay, stone, gravels, webs, insects, rodent hair and excreta	Visual examination will detect these adulterants
	Metanil yellow (a non permitted coaltar colour)	Take 5 gms of the sample with 5 ml. of water in a test tube and add a few drops of concentrated Hydrochloric acid. A pink colour shows presence o-Metanil yellow
Atta, Maida Suji (Rawa)	Sand, soil, insects, webs, lumps. rodent hair and excrete	These can be identified by visual examination.
	Iron filings	By moving a magnet through the sample, iron filings can be separated.
Bajra	Ergot infested Bajra.	Swollen and black Ergot infested grains will turn light in weight and will float also in water
Sago	Sand or talcum	Put a little quantity of sago in mouth, it will have a gritty feel, if adulterated. Burn the sago, if pure, it will swell and leave hardly any ash. Adulterated sago

		will leave behind appreciable quantity of ash.
Besan	Metanil Yellow	Take ½ teaspoon of the besan in a test tube. Pour 3 ml of alcohol in the test tube. Mix up the contents thoroughly by shaking the test tube. Add 10 drops of hydrochloric acid it. A pink colouration indicates presence of metanil yellow in the gram powder.
	Khesari Flour	Add 50 ml of dilute Hydrochloric acid to 10 gins. of sample and keep on simmering water for about 15 minutes. The pink colour, if developed, indicates, the presence of Khesari flour
Pulses	Lead Chromate	Shake 5 gm. of pulse with 5 ml. of water and add a few drops of HCl. Pink colour indicates Lead Chromate.
Whole spices	Dirt, dust, straw, insect, damaged seeds, other seeds, rodent hair and excrete	These can be examined visually
Black pepper	Papaya seeds	Papaya seeds can be separated out from pepper as they are shrunken, oval in shape and greenish brown or brownish black in colour.
	Light black	i) Float the sample of black pepper in alcohol (rectified spirit). The black pepper berries sink while the papaya seeds and light black pepper float.

	berries.	(ii) Press the berries with the help of fingers light peppers will break easily while black berries of pepper will not break.
	Coated with mineral oil	Black pepper coated with mineral oil gives Kerosene like smell.
Cloves	Volatile oil extracted (exhausted cloves)	Exhausted cloves can be identified by its small size and shrunken appearance. The characteristic pungent of genuine cloves is less pronounced in exhausted cloves
	Coated with mineral oil	Cloves coated with mineral oil gives kerosene like smell
Mustard seed	Argemone seed	Mustard seeds have a smooth surface The argemone seed have grainy and rough surface and are black and hence can be separated out by close examination. When Mustard seed is pressed inside it is yellow while for Argemone seed it is white
Powdered spices	Added starch	Add a few drops of tincture of Iodine or Iodine solution. Indication of blue colour shows the presence of starch.
	Chalk powder, yellow soap, stone	Take one gm of powdered spices in a test tube and add 5 ml of carbon- tetra-chloride solvent. Shake well and leave for some time. Impurities will settle at the bottom, while the spice powder will float on the surface.

powder.

	Common Salt	Taste for addition of common salt.
Turmeric powder	Coloured saw dust	Take a tea spoon full of turmeric powder in a test tube. Add a few drops of concentrated Hydrochloric acid. Instant appearance of pink colour which disappears on dilution with water shows the presence of turmeric If the colour persists, metanil yellow (an artificial colour) a now permitted coal tar colour is present.
	Chalk powder or yellow soap stone powder	Take a small quantity of turmeric powder in a test tube containing small quantity of water. Add a few drops of concentrated Hydrochloric acid, effervescence (give off bubbles) will indicate the presence of chalk or yellow soap stone powder
Turmeric powder	Starch of maize, wheat, tapioca, rice	A microscopic study reveals that only pure turmeric is yellow coloured, big in size and has an angular structure. While foreign/added starches are colourless and small in size as compared to pure turmeric starch.
Turmeric whole	Lead chromate	Appears to be bright in colour which leaves colour immediately in water.
Chillie powder	Brick powder, salt powder or talc. powder	Take a tea spoon full of chillies powder in a glass of water. Coloured water extract will show the presence of artificial colour. Any grittiness that may be felt on rubbing the sediment at the bottom of glass confirms the presence of brick powder/sand, soapy and smooth touch of the white residue at the

	bottom indicates the presence of soap stone. To a little powder of chilli add small amount of conc HCl and mix to the consistency of paste, dip the rear end of the match stick into the paste and hold over the flame, brick red flame colour due to the presence of calcium slats in brick powder.
Artificial colours	Sprinkle the chilli powder on a glass of water. Artificial colorants descend as coloured streaks.
Oil soluble coal tar colour	Take 2 gms of the sample in a test tube, add few ml of solvent ether and shake. Decant ether layer into a test tube containing 2 ml of dilute Hydrochloric acid (1 ml HCL plus 1 ml of water). Shake it, the lower acid layer will be coloured distinct pink to red indicating presence of oil soluble colour
Water soluble synthetic colour	Water soluble artificial colour can be detected by sprinkling a small quantity of chillies or turmeric powder on the surface of water contained in a glass tumbler. The water soluble colour will immediately start descending in colour streaks
Sudan III	Take 1 g of suspected chilli powder in a test tube and add 2ml of hexane to it shake well. Lt it settles for some time and decant the clear solution to another test tube. Ad 2 ml of aceto- nitrile reagent and shake well. The appearance of red colour in the lower aceto- nitrile layer is an indication of the presence of Sudan III.
Sawdust	Sprinkle chilli powder on the Sawdust will float on water and added surface of water

(Chilli powder)

Food	Adulterant	Test
Chilli powder	Rhodamine B	Take ¼ the teaspoon of the red chilli powder in a test tube. Add 3 ml of distilled water in it, and 10 drops of carbon- tetra- chloride. Vigorously shake the tube to mix up the contents. The red colour will disappear as the result of shake, and if the red colour reappears with the addition of a drop of hydrochloric acid, the adulteration of rhodamine B colour in the chilli powder is positive
Asafoetida (Hing)	Soap stone or other earthy mailer	Shake little portion of the sample with water and allow to settle. Soap stone or other earthy mailer will settle down at the bottom.
	Starch	Add tincture of iodine, appearance of blue colour shows the presence of starch.
	Foreign resin	Burn on a spoon, if the sample burns like camphor, it indicates the sample is pure.
Spices	Powdered bran and saw dust	Sprinkle on water surface. Powdered bran and sawdust float on the surface.
Cinnamon	Cassia bark	Cinnamon barks are very thin and can be rolled. It can be rolled around a pencil or pen. It has a distinct smell. Whereas cassia ark comprise of several layers in between the rough outer and inner most smooth layers. On examination of the ark loosely, a clear distinction can be made.
Cumin seeds	Grass seeds coloured	Rub the cumin seeds on palms. If palms turn black adulteration is indicated.

	with charcoal dust	
Green chilli, and green vegetables.	Malachite green	Take a cotton piece soaked in liquid paraffin and rub the outer green surface of a small part of green vegetable. If the cotton turns, green, we can say the vegetable is adulterated with malachite green.
Green peas Artificially coloured	Coloured	Take a little amount of green peas in a 250 ml beaker add water to it and mix well. Let it stand for half an hour. Clear separation of colour in water indicates adulteration.
Saffron	Dried tendrils of maizecob	Genuine saffron will not break easily like artificial. Artificial saffron is prepared by soaking maize cob in sugar and colouring it with coal tar colour. Genuine saffron will not break easily like spurious saffron. The colour dissolves in water if artificially coloured. A bit of pure saffron when allowed to dissolved in water will continue to give its saffron colour so long as it lasts
Powdered food stuff like betelnut spices, etc.	Sand, dirt, earth, gritty matter	Take a little amount of the sample in a test tube and add 10 ml of carbon tetrachloride to it. Shake well and allow standing for 5 minutes. Sand, dirt, earth, gritty matters, etc. will settle down at the bottom of the test tube.
Common salt	White powdered	Stir a spoonful of sample of salt in a glass of water. The presence of chalk will make solution white and other insoluble impurities will settle down.

Iodized salt	Common salt	Cut a piece of potato, add salt and wait minute and add two drops of lemon juice. If iodized salt blue colour will develop. In case of common salt, there will be no blue colour.
Tea leaves	Exhausted tea	Take a filter paper and spread a few tea leaves. Sprinkle with water to wet the filter paper. If coal tar colour is present it would immediately stain the filter paper. Wash the filter paper under tap water and observe the stains against light Spread a little slaked lime on white porcelain tile or glass plate; sprinkle a little tea dust on the lime. Red, orange or other shades of colour spreading on the lime will show the presence of coal tar colour. In case of genuine tea, there will be only a slight greenish yellow colour due to chlorophyll, which appear after some time
	Iron fillings	By moving a magnet through the sample, iron filling can be separated.
	Cereal starch	Take ¼ teaspoon of coffee powder in a test tube and add 3 ml of distilled water in it. Light a spirit lamp and heat the contents to colorize. Add about 33 ml of potassium per magnate solution and muriatic acid (1:1) to decolorize the mixture. The formation of blue colour in the mixture, when adding a drop of 1% aqueous solution of iodine indicates adulteration with starch.
Coffee powder	Chicory	Gently sprinkle the coffee powder sample on the surface of water in a glass. The coffee floats over the water but chicory begins to sink down within a few seconds. The falling chicory powder particles leave behind them a trail of colour, due to large amount of caramel

Coffee Powder	Tamarind seed/ date	Sprinkle the suspected coffee powder on white filter/blotting paper and spray I per cent sodium carbonate solution on it. Tamarind and date seed powder will, if present, stain blotting paper/filter paper red.
	Corched persimmon	Take 1 tsp of coffee powder and spread it on a moisturized blotting paper. Pour 3 ml of 2% aqueous solution of sodium carbonate slowly and carefully on it. A red coloration indicates the presence of the powder of scorched persimmon stones in the coffee powder.
Supari Pan Masala	Colour	Colour dissolves in water immediately.
	Saccharin	Saccharin gives excessive and lingering sweet taste and leaves bitter taste at the end.
Catachu powder	Chalk	Chalk gives effervescence (gives off bubbles) with concentrated Hydrochloric acid
Processed foods, sweets and syrups	Rhoda mine B	If this chemical colour is present in the food, it is very easy to detect. Because it shines very brightly under the sun. Also it can be detected by a more precise method. Take ½ teaspoon of the sample in a test tube. Pour 3 ml of carbon tetra chloride and shake the test tube to mix the contents. The mixture turns colourless and addition of a drop of hydrochloric acid brings the colour back, when food contains Rhodamine b colour.
Lemonade	Mineral acid	Pour 2 drops of the lemonade soda on a metanil yellow paper- strip. A violet

soda		colouration indicates the presence of mineral acid in aerated water. The colour impression gets retained even after drying the paper (you can prepare metanil yellow paper strips by soaking filter paper strips in 0.1 % aqueous solution and then drying the paper – strips).
Sweet Potato	Rhodamine B colour	Take a cotton piece soaked in liquid paraffin, and rub the outer red surface of the sweet potato. If the cotton absorb colour, it indicates the use of Rhodamine B colours on the outer surface of the sweet potato.
Pulses	Lead Chromate	Shake 5 gm. of pulse with 5 ml. of water and add a few drops of HCl. Pink colour indicates Lead Chromate.
Silver leaves	Aluminium leaves	(i) On ignition, genuine silver leaves burn away completely, leaving glistering white spherical ball of the same mass whereas aluminium leaves are reduced to ashes of dark grey blackish colour.
		(ii) Take silver leaves in test tube, add diluted Hydrochloric acid. Appearance of turbidity to white precipitate indicates the presence of silver leaves. Aluminium leaves do not give any turbidity or precipitate.
		(iii) Take a small portion of metal leaves and add a few drops of concentrated Nitric acid. Silver leaves will completely dissolve whereas aluminium leaves will remain undisclosed.
Vinegar	Mineral Acid	Test with the Metanil yellow indicator paper, in case, the colour changes from yellow to pink, mineral acid is present.

ANALYTICAL PROCEDURES FOR DETECTING 210 FOOD ADDITIVES (Source: The Joint FAO/WHO Expert Committee on Food Additives (JECFA), Food and Agriculture Organization of the United Nation Organization)

Food additive	Simple Analytical Procedure
Acesulfame potassium	Dissolve about 0.15 g of the dried sample (dissolution may be slow), accurately weighed, in 50.0 ml glacial acetic acid and titrate potentiometrically with 0.1 N perchloric acid, or add two drops of crystal violet TS and titrate with 0.1 N perchloric acid, to a blue-green end-point which persists for at least 30 sec. Perform a blank determination and make any necessary correction. Each ml of 0.1 N perchloric acid is equivalent to 20.12 mg of C4H4KNO4S.
Acetic acid ethyl ester	Transfer about 1.5 g, accurately weighed in a tared stoppered weighing bottle, to a suitable flask, add 50.0 ml of 0.5N sodium hydroxide, and reflux on a steam bath for 1 h. Allow to cool, add phenolphthalein TS and titrate the excess sodium hydroxide with 0.5N hydrochloric acid. Perform a blank determination, and make any necessary correction. Each ml of 0.5N sodium hydroxide is equivalent to 44.06 mg of $C_4H_8O_2$.
Adipic acid	Transfer about 3 g of the sample, accurately weighed, into a 250-ml conical flask, add 50 ml of methanol and dissolve the sample by warming gently on a steam bath. Cool, add phenolphthalein TS, and titrate with 1 N sodium hydroxide. Perform a blank determination and make any necessary correction. Each ml of 1 N sodium hydroxide is equivalent to 73.07 mg of $C_6H_{10}O_4$.
Agar	Threshold gel concentration: Prepare serial dilutions of the sample with known solids content

(0.15%, 0.20%, 0.25%, etc.) and place in tubes, 150 mm long by 16 mm internal diameter, stoppered at both ends. Cool for 1 h at 20-25^0. Allow cylinders of gel to slide from the tubes to a level surface. The lowest concentration of gel that resists gravity without rupture for 5-30 sec is the threshold concentration of the sample.

Aluminium ammonium sulfate

Weigh accurately about 1 g of the sample, dissolve in 50 ml of water, add 50 ml of 0.05 M disodium EDTA and 20 ml of pH 4.5 buffer solution (77.1 g of ammonium acetate and 57 ml of glacial acetic acid in 1000 ml of solution), and boil gently for 5 min. Cool, and add 50 ml of ethanol and 2 ml of dithizone TS. Titrate with 0.05 M zinc sulfate to a bright rose-pink colour, and perform a blank determination. Each ml of 0.05 M disodium EDTA is equivalent to 22.67 mg of AlNH4(SO4)2 · 12H2O.

Aluminium potassium sulfate

a) Dodecahydrate form

Weigh accurately about 1 g of the sample, dissolve in 50 ml of water, add 50.0 ml of 0.05 M disodium ethylenediamine-tetraacetate, and boil gently for 5 min. Cool, and with continuous stirring add in the order given: 20 ml of pH 4.5 buffer solution (77.1 g of ammonium acetate and 57 ml of glacial acetic acid in 1000 ml), 50 ml of ethanol, and 2 ml of dithizone TS. Titrate with 0.05 M zinc sulfate to a bright rose-pink colour, and perform a blank determination. Each ml of 0.05 M disodium ethylenediamine tetraacetate is equivalent to 23.72 mg of AlK(SO$_4$)$_2$ · 12H$_2$O.

b) Anhydrous form

Weigh accurately about 0.8 g of powdered Aluminium Potassium Sulfate, previously dried at 200° for 4 hours. Add 100 ml of water, dissolve by heating in a water bath and shaking, filter, and wash the insoluble residue thoroughly with water. Combine the filtrate and the washings, add water to make exactly 200 ml. Measure 25 ml of this solution, add 50 ml of 0.01mol/l EDTA, and heat to boiling. Cool and add 7 ml of 14% sodium acetate solution and 85 ml of absolute ethanol. Titrate the excess EDTA with 0.01 mol/l zinc acetate using 3 drops of xylenol orange TS as indicator. The end point is when the yellow colour of the solution changes to red.

Aluminium powder	Wash a small sample in hexane, repeating to remove traces of any associated oil or fatty acid. Transfer about 0.2 g of the sample, accurately weighed, to a 500 ml flask fitted with a rubber stopper carrying a 150 ml separating funnel, an inlet tube connected to a cylinder of carbon dioxide and an outlet tube dipping into a water-trap. Add 60 ml of freshly boiled and cooled water and disperse the sample, replace the air by carbon dioxide and add, by the separating funnel, 100 ml of a solution containing 56 g of ferric ammonium sulfate and 7.5 ml of sulfuric acid in freshly boiled and cooled water. While maintaining an atmosphere of carbon dioxide in the flask, heat to boiling and boil for 5 min. After the sample has dissolved, cool rapidly to 20o, and dilute to 250 ml with freshly boiled and cooled water. To 50 ml of this solution, add 15 ml of phosphoric acid and titrate with 0.1 N potassium permanganate. 1 ml of 0.1 N potassium permanganate is equivalent to 0.8994 mg of Al.
Aluminium sulfate	Weigh accurately about 4 g of the sample, transfer into a 250-ml volumetric flask. Dissolve in water, dilute to volume with water, and mix. Pipet 10 ml of this solution into a 250-ml beaker, add 25.0 ml of 0.05 M disodium ethylenediaminetetra-acetate, and boil gently for 5 min. Cool, and with continuous stirring add in the order given: 20 ml of pH 4.5 buffer solution (77.1 g of ammonium acetate and 57 ml of glacial acid in 1000 ml), 50 ml of ethanol, and 2 ml of dithizone TS. Titrate with 0.05 M zinc sulfate until the colour changes from green-violet to rose-pink, and perform a blank determination, substituting 10 ml of water for the sample. Each ml of 0.05 M

disodium ethylenediaminetetraacetate is equivalent to 8.553 mg of $Al_2(SO_4)_3$.

Ammonium bicarbonate	Place about 10 ml of water in a weighing bottle, tare the bottle and its contents, add about 2 g of the sample and weigh accurately. Transfer the contents of the bottle to a 250-ml flask and slowly add, with mixing, 50 ml of 1 N sulfuric acid. When solution has been effected, wash down the sides of the flask, add methyl orange TS, and titrate the excess acid with 1 N sodium hydroxide. Each ml of 1 N sulfuric acid is equivalent to 79.06 mg of NH_4HCO_3.
Ammonium dihydrogen phosphate	Dissolve about 500 mg of the sample, accurately weighed, in 50 ml of water, and titrate to a pH of 8.0 with 0.1 N sodium hydroxide. Each ml of 0.1 N sodium hydroxide is equivalent to 11.50 mg of $NH_4H_2PO_4$.
Ammonium chloride	Dry about 0.2 g of the sample over silica gel for 4 h, weigh accurately, and dissolve it in about 40 ml of water in a glass-stoppered flask. Add, while agitating, 3 ml of nitric acid, 5 ml of nitrobenzene, 50.0 ml of 0.1 N silver nitrate, shake vigorously, then add 2 ml of ferric ammonium sulfate TS, and titrate the excess silver nitrate with 0.1 N ammonium thiocyanate. Each ml of 0.1 N silver nitrate is equivalent to 5.349 mg of NH_4Cl.
Ammonium hydrogen carbonate	Place about 10 ml of water in a weighing bottle, tare the bottle and its contents, add about 2 g of the sample and weigh accurately. Transfer the contents of the bottle to a 250-ml flask and slowly add, with mixing, 50 ml of 1 N sulfuric acid. When solution has been effected, wash down the sides of the flask, add methyl orange TS, and titrate the excess acid with 1 N sodium hydroxide. Each ml of 1 N sulfuric acid is equivalent to 79.06 mg of NH_4HCO_3.
Ammonium	Transfer about 1 g of the sample, accurately weighed, into a 250 ml glass-stoppered Erlenmeyer

ferric citrate

flask, and dissolve in 25 ml of water and 5 ml of hydrochloric acid. Add 4 g of potassium iodide, stopper, and allow to stand protected from light for 15 min. Add 100 ml of water, and titrate the liberated iodine with 0.1 N sodium thiosulfate, using starch TS as the indicator. Perform a blank determination and make any necessary correction. Each ml of 0.1 N sodium thiosulfate is equivalent to 5.585 mg of iron (Fe).

Ammonium glutamate

Dissolve about 200 mg of the sample, previously dried and weighed accurately, in 6 ml of formic acid, and add 100 ml of glacial acetic acid. Titrate with 0.1 N perchloric acid determining the end-point potentiometrically. Run a blank determination in the same manner and correct for the blank. Each ml of 0.1 N perchloric acid is equivalent to 9.106 mg of $C_5H_{12}N_2O_4 \cdot H_2O$.

Ammonium carbonate

Place about 10 ml of water in a weighing bottle, tare the bottle and its contents, add about 2 g of the sample and weigh accurately. Transfer the contents of the bottle to a 250-ml flask and slowly add, with mixing, 50 ml of 1 N sulfuric acid. When solution has been effected, wash down the sides of the flask, add methyl orange TS, and titrate the excess acid with 1 N sodium hydroxide. Each ml of 1 N sulfuric acid is equivalent to 17.03 mg of NH_3.

Ammonia solution

Tare accurately a 125-ml glass-stoppered conical flask containing 35.0 ml of 1N sulfuric acid. Cool the sample in the original bottle to 10° or lower. Partially fill a 10-ml graduated pipet from near the bottom (do not use vacuum for drawing up the sample). Wipe off any liquid adhering to the outside of the pipet and discard the first ml. Hold the pipet just above the surface of the acid and transfer 2 ml into the flask, leaving at least 1 ml in the pipet. Stopper the flask, mix and weigh again to obtain the weight of the sample. Add methyl red TS and titrate the excess acid with 1N sodium hydroxide. Subtract the excess sulfuric acid from the total sulfuric acid (35.0 ml) to find the ml used to neutralize the sample. Each ml of 1N sulfuric acid used to neutralize the ammonia is equivalent to 17.03 mg of NH_3.

Triammonium citrate	Dissolve about 3.5 g of the sample, accurately weighed, in 50 ml of water, add 50 ml of 1 N sodium hydroxide, boil for 15 min or until ammonia ceases to be evolved, add sufficient 1 N sulfuric acid to make the solution acid to phenolphthalein TS, boil for 5 min, cool, and titrate with 1 N sodium hydroxide, using phenolphthalein TS as an indicator. Each ml of 1 N sodium hydroxide is equivalent to 81.07 mg of $C_6H_{17}N_3O_7$.
Triammonium citrate	Dissolve about 3.5 g of the sample, accurately weighed, in 50 ml of water, add 50 ml of 1 N sodium hydroxide, boil for 15 min or until ammonia ceases to be evolved, add sufficient 1 N sulfuric acid to make the solution acid to phenolphthalein TS, boil for 5 min, cool, and titrate with 1 N sodium hydroxide, using phenolphthalein TS as an indicator. Each ml of 1 N sodium hydroxide is equivalent to 81.07 mg of $C_6H_{17}N_3O_7$.
Argon	After determination of the total content of water, hydrogen, oxygen and nitrogen the balance consists of argon
Aspartame	Weigh accurately about 150 mg of the sample, previously dried at 105o for 4h dissolve in 35 ml of dimethylformamide, add 5 drops of thymol blue TS, and titrate with a microburette to a dark blue end-point with 0.1 N lithium methoxide. Perform a blank determination and make any necessary correction. Each ml of 0.1 N lithium methoxide is equivalent to 29.43 mg of C14H18N2O5.
	Caution: Protect the solution from absorption of carbon dioxide and moisture by covering the titration vessel with aluminium foil while dissolving the sample and during the titration.

Aspartame-acesulfame	Weigh accurately 100 to 150 mg of sample and dissolve it in 50-ml methanol. Titrate with the standardized 0.1 M tetrabutylammonium hydroxide. Determine the volume (ml) of the standard solution needed to reach the first (V1) and second (V2) equivalency points. Perform a blank titration on the methanol.
Beetred	Dissolve a quantity of Beet Red accurately weighed in buffer TS (pH 5) and dilute to a suitable volume with the buffer solution (V ml in total); the maximum absorption shall be within the range of 0.2 to 0.8. Centrifuge the solution if necessary, and measure the absorption, correcting for a blank composed of Buffer TS (pH 5). The colour content is calculated on the basis of the maximum absorption A (at about 530 nm), using the specific absorbance for betanine, A (1%, 1 cm) = 1120.
Benzoyl peroxide	Dissolve about 250 mg of the sample, accurately weighed, in 15 ml of acetone in a 100-ml glass-stoppered bottle. Add 3 ml of 50% (w/v) potassium iodide solution and swirl for 1 min. Titrate immediately with 0.1 N sodium thiosulfate (without addition of starch as an indicator). Each ml of 0.1 N sodium thiosulfate is equivalent to 12.11 mg of $C_{14}H_{10}O_4$.
Benzyl alcohol	Weigh accurately about 1 g of the sample, proceed as directed under the method for *Hydroxyl Value* and calculate the percentage of benzyl alcohol by the formula % w/w = W (S) x N x 10.814 C (B + W x A), where, A is ml of KOH solution required for the free acid determination; B is ml of KOH solution required for the reagent blank; C is weight of the sample used for the free acid determination; S is ml of KOH solution required for the titration of the acetylated sample; W is the weight of the sample used for acetylation; and N is the normality of the ethanolic KOH solution.

Blackcurrant extract

Distil 10 g of the sample with 100 ml of water and 25 ml of 30% phosphoric acid solution in a distilling flask with the Wagner tube. In an absorption flask, place 25 ml of 2% lead acetate solution previously prepared. Insert the lower end of the condenser into the lead acetate solution in the absorption flask. Distil until the liquid in the absorption flask reaches about 100 ml and rinse the end of the condenser with a little amount of water. To the distilled solution add 5 ml of hydrochloric acid and 1 ml of starch TS, and titrate with 0.01 N iodine. Each ml of 0.01 N iodine is equivalent to 0.3203 mg of SO_2.

Calcium aluminium silicate

Silicon dioxide:

Transfer 500 mg of the sample, previously dried at 105^0C for 2 h and weighed accurately, into a 250 ml beaker. Wash the walls of beaker with a few ml of water, then add 30 ml of 72% perchloric acid and 15 ml of hydrochloric acid. Heat on a hot-plate until dense white fumes appear. Let cool. Add 15 ml of hydrochloric acid and reheat until dense fumes appear. Let cool, add 70 ml of water, and filter through Whatman No. 40 filter paper or equivalent. Wash the paper and the precipitate with hot water to remove perchloric acid. Then transfer the paper and the precipitate to a tared platinum crucible, and ignite at 900o to constant weight. Moisten the residue with a few drops of water, then add 15 ml of hydrochloric acid and 8 drops of sulfuric acid, and heat on a hotplate until white fumes of sulfur trioxide appear. Let cool. Add 5 ml of water, 10 ml of hydrofluoric acid (warning: toxic, corrosive, must not contact skin; work under fumehood) and 3 drops of sulfuric acid, then evaporate to dryness on a hotplate. Then ignite at 900^0C to constant weight. The weight loss after the sulfur trioxide no longer appear. Then ignite at 900^0C to constant weight. The weight loss after the addition of hydrofluoric acid represents the weight of SiO2 in the sample taken.

Aluminium oxide:

Fuse the residue obtained in the silicon dioxide determination with 2 g of potassium pyrosulfate for 5 min. Cool, dissolve the fusion in water, and dilute to 250 ml in a volumetric flask. Transfer 100 ml of the solution into a 600 mlbeaker, add 100 ml of water and 5 drops of bromotymol blue TS, and heat to as low boil. Add ammonium hydroxide, dropwise, until a blue colour appear,then boil the solution for 5 min. to expel the excess ammonia. Filter through Whatman No. 41, or equivalent filter paper, and wash the precipitate with six portions of a 1-in-50 hot ammonium chloride solution. Transfer the filter and precipitate into a tared platinum crucible, char the paper, and ignite over a Meker burner to constant weight. The weight of the residue, corrected for the ash content of the filter paper and multiplied by 2.5, represents the weight of Al2O3 in the original sample.

Calcium oxide:

To the combined filtrate and washings retained in the silicon dioxide determination, add, while stirring, about 30 ml of 0.05 M disodium ethylendiaminetetraacetate from a 50-ml burette. Then add 15 ml of sodium hydroxide TS and 300 g of hydroxynaphthol blue indicator, and continue the titration to a blue end point. Each ml of 0.05 M disodium ethylenediaminetetraacetate is equivalent to 2.804 mg of CaO.

Sodium oxide:

Transfer about 500 mg of the sample, previously dried at 105o for 2 h, and accurately weighed, into a tared platinum dish, and moisten with 8 to 10 drops of water. Add 25 ml of 70% perchloric acid and 10 ml of hydrofluoric acid (warning: toxic, corrosive, must not contact skin; work under fumehood) and heat on a hot plate in a hood until dense white fumes of perchloric acid appear. Add 10 ml of hydrofluoric acid (warning: toxic, corrosive, must not contact skin; work under fumehood), heat again to dense white fumes, and dissolve the residue in sufficient water to make 250 ml. Set a suitable flame photometer to a wavelength of 589 nm. Adjust the instrument to zero transmittance against water, then adjust it to 100% transmittance with a standard solution containing 200 µg of sodium, in the form of the chloride, per ml. Read the percent transmittance of three other standard solutions containing 50, 100 and 150 µg each of sodium per ml, and plot the standard curve as % transmittance *vs.* concentration of sodium. Place a portion of the sample solution in the photometer, read the percent transmittance in the same manner, and by reference to the standard curve determine the concentration (C) of sodium, in T per ml in the sample solution. Calculate the quantity, in mg, of Na2O in the sample taken by the formula: where F is the quantity of sodium oxide equivalent to any sodium sulfate present in the sample, is found as follows:

Correction for sodium sulfate content:

Weigh accurately 12.5 g of the sample, previously dried at 105o for 2 h and stir it with 240 ml of water for at least 5 min. with a high speed mixer. Transfer the mixture into a 250-ml graduated cylinder, and wash the mixer container with water, adding the washings to the cylinder to make 250 ml. Stopper the cylinder, invert it several times to mix the sample, and determine the conductivity of the slurry using a suitable conductance bridge assembly. By means of a standard curve, obtained from solutions containing 50, 100, 200 and 500 mg of sodium sulfate per 100 ml, determine the concentration (C), in mg per 100 ml, of sodium sulfate in the sample slurry.

Calcium5'

Weigh accurately about 500 mg of the sample, dissolve in and make to 1,000 ml with 0.01 N hydrochloric acid. Take 10.0 ml of this solution and dilute with 0.01 N hydrochloric acid to 250

guanylate

ml. Determine the absorbance A of the solution in a 1-cm cell at the wave length of 260 nm using 0.01 N hydrochloric acid as the reference.

Calcium 5'-inosinate

Weigh accurately about 500 mg of the sample, dissolve in and make to 1,000 ml with 0.01 N hydrochloric acid. Take 10.0 ml of this solution and dilute with 0.01 N hydrochloric acid to 250 ml. Determine the absorbance A of the solution in a 1-cm cell at the wave length of 260 nm using 0.01 N hydrochloric acid as the reference.

Calcium acetate

Calcium content:

Dissolve in a beaker 2.5 g of the sample, weighed to the nearest mg, in 5 ml of hot dilute hydrochloric acid TS. Cool, transfer to a 250-ml volumetric flask, dilute to volume with water, and mix. Transfer 50 ml of the solution to a 400-ml beaker, add 100 ml of water, 25 ml of sodium hydroxide TS, 40 mg of murexide indicator preparation (an alternative indicator is hydroxynaphtol blue, of which 0.25 g is used - in this case the naphthol green TS is omitted), and 3 ml of naphthol green TS. Titrate with 0.05 M disodium ethylenediamine-tetraacetate until the solution is deep blue in colour. Each ml of 0.05 M disodium ethylenediaminetetraacetate is equivalent to 7.909 mg of $C_4H_6CaO_4$.

Acid content:

Half fill a chromatographic column (1.5 cm in diameter, 20 cm long) with a strong cation-exchange resin (Amberlite IR 120, Amberlite IR 100, Duolit C III, Dorvex 50, Lewatit KS, Ion Exchanger I Merck). Add 0.1 N hydrochloric acid through the top of the column, with the outflow orifice closed until the resin is completely covered and let stand 1-2 h. Drain the acid and rinse the column with water (about 1 liter) until 20 ml of eluate forms a red colour, when one drop each of

0.02 N sodium hydroxide and phenolphthalein TS is added. Weigh, to the nearest mg, 0.05 g of the sample, previously dried at 155 ° to constant weight, into a flask. Dissolve in 15 ml of water and pour slowly on to the column. Wash the flask and the column with about 200 ml of water and collect the total filtrate in a conical flask. Add two drops of phenolphthalein TS and titrate with 0.1 N sodium hydroxide using a microburette. Each ml of 0.1 N sodium hydroxide is equivalent to 7.909 mg of C4H6CaO4.

Calcium alginate

Proceed as directed under Carbon Dioxide Determination by Decarboxylation in the General Methods (See Volume 4). Each ml of 0.25 N sodium hydroxide consumed is equivalent to 5.5 mg of carbon dioxide (CO_2) or 27.38 mg of calcium alginate (equivalent weight 219).

Calcium ascorbate

Weigh accurately about 0.4 g of the sample into a 250 ml flask and add 50 ml of carbone dioxide free water. Immediately titrate with 0.1 N iodine, adding a few drops of starch TS as indicator as the end point is approached. Each ml of 0.1 N iodine is equivalent to 10.66 mg of C12H14O12Ca . 2H2O.

Calcium benzoate

Weigh accurately 0.6 g of the dried sample, dissolve in a mixture of 20 ml of water and 2 ml of dilute hydrochloric acid TS, and dilute to 100 ml with water. While stirring (preferably with a magnetic stirrer) add about 30 ml of 0.05M disodium ethylenediaminetetraacetate from a 50-ml buret, then add 15 ml of sodium hydroxide TS, 40 mg of murexide indicator preparation (an alternative indicator is hydroxynaphthol blue, of which 0.25 g is used - in this case the naphthol green TS is omitted) and 3 ml of naphthol green TS, and continue the titration until the solution is deep blue in colour. Each ml of 0.05M disodium ethylenediamine tetraacetate is equivalent to 14.116 mg of $C_{14}H_{10}CaO_4$.

Calcium carbonate	Weigh, to the nearest 0.1 mg, 200 mg of the dried sample. Transfer into a 400-ml beaker, add 10 ml of water and swirl to form a slurry. Cover the beaker with a watch glass and introduce 2 ml of dilute hydrochloric acid TS from a pipette inserted between the lip of the beaker and the edge of the watch glass. Swirl the contents of the beaker to dissolve the sample. Wash down the sides of the beaker, the outer surface of the pipette and the watch glass, and dilute to about 100 ml with water.
Calcium chloride	Weigh accurately about 1 g of anhydrous calcium chloride, or the corresponding weight of a hydrate, transfer to a 250-ml beaker, and dissolve in a mixture of 100 ml of water and 5 ml of dilute hydrochloric acid TS. Transfer the solution to a 250-ml volumetric flask, dilute with water to volume and mix. Pipet 50 ml of the solution into a suitable container, add 100 ml of water, 15 ml of sodium hydroxide TS, 40 mg of murexide indicator (amm. purpurate) and 3 ml of naphthol green TS, and titrate with 0.05 M disodium ethylenediaminetetra-acetate until the solution is deep blue in colour. Each ml of 0.05 M disodium ethylenediaminetetraacetate is equivalent to 5.55 mg of $CaCl_2$; 7.35 mg of $CaCl_2 \cdot 2H_2O$; or 10.95 mg of $CaCl_2 \cdot 6H_2O$
Calcium citrate	Weigh accurately about 350 mg of the sample, previously dried at 150o for 4 h, dissolve in a mixture of 10 ml of water and 2 ml of dilute hydrochloric acid TS, and dilute to about 100 ml with water. While stirring (preferably with a magnetic stirrer) add about 30 ml of 0.05 M disodium ethylenediaminetetraacetate from a 50-ml buret, then add 15 ml of sodium hydroxide TS and 300 mg of hydroxynaphthol blue indicator, and continue the titration to a blue endpoint. Each ml of 0.05 M disodium ethylenediaminetetraacetate is equivalent to 8.303 mg of C12H10Ca3O14.
Calcium	Dissolve about 0.4 g of the sample accurately weighed, in a mixture of 50 ml of water and 5 ml of dilute hydrochloric acid TS and titrate with 0.1M sodium nitrite. Add the last ml of titrant

cyclamate

dropwise until a blue colour is produced immediately when a glass rod dipped into the titrated solution is streaked on a piece of starch iodide test paper or alternatively the end point may be detected electro-metrically. When the titration is complete, the end-point is reproducible after the mixture has been allowed to stand for 1 min. Each ml of 0.1M sodium nitrite is equivalent to 19.83 mg of C12H24CaN2O6S2, calculated on the anhydrous basis.

Calcium dihydrogen diphosphate

Weigh accurately about 200 mg of the sample, dissolve in 25 ml of water and 10 ml of diluted nitric acid TS and boil for 30 min. Filter if necessary, and wash any precipitate, then Dissolve the precipitate by the addition of 1 ml diluted nitric acid TS. Adjust the temperature to about 50^0C, add 75 ml of ammonium molybdate TS, and maintain the temperature at about 50^0C for 30 min, stirring occasionally. Allow to stand for 16 h or overnight at room temperature. Decant the supernate, through a filter paper, wash the precipitate once or twice with water by decantation using 30 to 40 ml each time, and pour the washings through the same filter. Transfer the precipitate to the same filter, and wash with potassium nitrate solution (1 in 100) until the filtrate is no longer acid to litmus paper. Transfer the precipitate with filter paper to the original precipitation vessel, add 50.0 ml of 1N sodium hydroxide, agitate until the precipitate is dissolved, add 3 drops of phenolphthalein TS and titrate the excess alkali with 1N sulfuric acid. Each ml of 1N sodium hydroxide consumed is equivalent to 3.088 mg of P_2O_5.

Calcium dihydrogen phosphate

Weigh accurately a portion of the sample equivalent to about 475 mg of the anhydrous salt and dissolve it in 10 ml of hydrochloric acid TS. Add a few drops of methyl orange TS, and boil for 5 min, keeping the volume and pH of the solution constant during the boiling period by adding hydrochloric acid or water, if necessary. Add 2 drops of methyl red TS and 30 ml of ammonium oxalate TS, then add dropwise, with constant stirring, a mixture of equal volumes of 6N ammonia solution and water until the pink colour of the indicator just disappears. Digest on a steam bath for 30 min, cool to room temperature, allow the precipitate to settle, and filter the supernatant liquid through a sintered-glass crucible, using gentle suction. Wash the precipitate in the beaker with about 30 ml of cold (below 200) wash solution, prepared by diluting 10 ml of ammonium oxalate TS to 1000 ml. Allow the precipitate to settle, and pour the supernatant liquid through the filter. Repeat this washing by decantation three more times. Using the wash solution, transfer the precipitate as completely as possible to the filter. Finally, wash the beaker and the filter with two 10 ml portions of cold (below 200C) water. Place the sintered-glass crucible in the beaker, and add 100 ml of water and 50 ml of cold dilute sulfuric acid (1 in 6). Add from a buret 35 ml of 0.1N potassium permanganate, and stir until the colour disappears. Heat to about 700, and complete the titration with 0.1N potassium permanganate. Each ml of 0.1N potassium permanganate is equivalent to 2.004 mg of Ca.

Calcium di-l-glutamate

Dissolve about 250 mg of the sample, previously dried and weighed accurately, in 6 ml of formic acid, and add 100 ml of glacial acetic acid. Titrate with 0.1 N perchloric acid determining the end-point potentiometrically. Run a blank determination in the same manner and correct for the blank. Each ml of 0.1 N perchloric acid is equivalent to 8.308 mg of $C_{10}H_{16}CaN_2O_8$. Calculate the content on the anhydrous basis.

Calcium disodium ethylenediaminetetraacetate

Transfer about 1.2 g of the sample, accurately weighed, into a 250-ml beaker, and dissolve in 75 ml of water. Add 25 ml of dilute acetic acid TS and 1.0 ml of diphenylcarbazone solution (1 g in 100 ml ethanol). Titrate slowly with 0.1 M mercuric nitrate (see below) to the first appearance of a purplish colour. Each ml of 0.1 M mercuric nitrate is equivalent to 37.43 mg of $C_{10}H_{12}CaN_2Na_2O_8$.

Mercuric nitrate solution: Dissolve about 35 g of mercuric nitrate $Hg(NO3)2 \cdot H2O$ in a mixture of 5 ml of nitric acid and 500 ml of water and dilute with water to 1000 ml. Standardize the solution as follows: Transfer an accurately measured volume of about 20 ml of the solution into an Erlenmeyer flask and add 2 ml of nitric acid and 2 ml of ferric ammonium sulfuric TS. Cool to below 20°C and titrate with 0.1 N ammonium thiocyanate to the first appearance of a permanent brownish colour. Calculate the molarity. (0.1 M = 32.46 g Hg(NO3)2 per litre).

Calcium dl-malate

Weigh accurately about 0.4 g of the sample, previously dried at 110o for 3 h, dissolve in a mixture of 10 ml of water and 2 ml of dilute hydrochloric acid TS, and dilute to about 100 ml with water. While stirring (preferably with a magnetic stirrer) add about 30 ml of 0.05 M disodium ethylenediaminetetraacetate from a 50-ml buret, then add 15 ml of sodium hydroxide TS and 300 mg of hydroxynaphtol blue indicator, and continue the titration to a blue endpoint. Each ml of 0.05 M disodium ethylenediaminetetraacetate is equivalent to 8.607 mg of C4H4CaO5.

Calcium hydrogen phosphate

Weigh accurately about 0.3 g of the sample, previously dried for 3 h at 200˚. Dissolve in 10 ml of dilute hydrochloric acid TS, add about 120 ml of water and a few drops of methyl orange TS, and boil for 5 min, keeping the volume and pH of the solution in the beaker constant during the boiling period by adding hydrochloric acid or water as necessary. Add 2 drops of methyl red TS and 30 ml of ammonium oxalate TS. Then add drop-wise, with constant stirring, a mixture of equal volumes of ammonia TS and water until the pink colour of the indicator just disappears. Digest on a steam bath for 30 min, cool to room temperature, allow the precipitate to settle, and filter the supernatant liquid through an asbestos mat in a Gooch crucible, using gentle suction. Swirl the precipitate in the beaker with about 30 ml of a cold (below 20o) wash solution prepared by diluting 10 ml of ammonium oxalate TS to 1000 ml. Allow the precipitate to settle, and pass the supernatant through the filter. Repeat this washing by decantation three more times. Using the wash solution, transfer the precipitate as completely as possible to the filter. Finally, wash the beaker and the filter with to 10 ml portions of cold (below 20o) water. Place the Gooch crucible in the beaker, and add 100 ml of water and 50 ml of cold dilute sulfuric acid (1 in 6). Add from a buret 35 ml of 0.1 N potassium permanganate, and stir until the colour disappears. Heat to about 70o, and complete the titration with 0.1 N potassium permanganate. Each ml of 0.1 N potassium permanganate is equivalent to 6.803 mg of $CaHPO_4$.

Calcium gluconate	Weigh accurately about 0.5 g of the dried sample and dissolve in 5 ml of dilute hydrochloric acid. Add 50 ml of water, 25 ml of sodium hydroxide TS and about 0.1 g of 2-hydroxy-1-(2'-hydroxy-4'-sulfo-1'-naphthylazo)-3-naphthoic acid. Titrate with 0.05 M EDTA immediately. At the end-point, the red colour changes completely to blue. Each ml of 0.05 M EDTA is equivalent to 22.42 mg of $C_{12}H_{22}CaO_{14} \cdot H_2O$.
Calcium hydroxide	Weigh accurately about 1.5 g of the sample, transfer to a beaker, and gradually add 30 ml of dilute hydrochloric acid TS. When solution is complete, transfer to a 500-ml volumetric flask, rinse the beaker thoroughly, adding the rinsings to the flask, dilute to volume with water, and mix. Pipet 50 ml of the solution into a suitable container and add 50 ml of water and 15 ml of sodium hydroxide TS, 40 mg of murexide indicator (amm. purpurate) and 3 ml of naphthol green TS, and titrate with 0.05 M disodium ethylenediaminetetraacetate until the solution is deep blue in colour. Each ml of 0.05 M disodium ethylenediaminetetraacetate is equivalent to 3.705 mg of $Ca(OH)_2$.
Calcium lactate	Dissolve about 350 mg of previously dried sample, accurately weighed, in 150 ml of water containing 2 ml of dilute hydrochloric acid TS. While stirring, preferably with a magnetic stirrer,

add about 30 ml of 0.05 M disodium ethylenediaminetetraacetate from a 50-ml buret. Then add 15 ml of sodium hydroxide TS and 300 mg of hydroxy-naphthol blue indicator, and continue the titration to a blue end-point. Each ml of 0.05 M disodium ethylenediaminetetraacetate is equivalent to 10.91 mg of $C_6H_{10}CaO_6$.

Calcium oxide
Ignite at approximately 800^0F about 1 g of the sample to constant weight, accurately weigh the residue and dissolve it in 20 ml of dilute hydrochloric acid TS. Cool the solution, dilute with water to 500 ml and mix. Pipet 50 ml of this solution into a suitable container and add 50 ml of water, then add 15 ml of sodium hydroxide TS, 40 mg of murexide indicator preparation and 3 ml of naphthol green TS, and titrate with 0.05 M disodium ethylenediamine tetraacetate until the solution is deep blue in colour. Each ml of 0.05 M disodium ethylenediamine tetraacetate is equivalent to 2.804 mg of CaO.

Calcium polyphosphate
Mix about 300 mg of the sample, accurately weighed, with 15 ml of nitric acid and 30 ml of water, boil for 30 min and dilute with water to about 100 ml. Heat at 60^0C, and add excess of ammonium molybdate TS, and heat at 50^0C for 30 min. Filter, and wash the precipitate with dilute nitric acid (1 in 36), followed by potassium nitrate solution (1 in 100) until the filtrate is no longer acid to litmus

Calcium propionate
Dissolve in a beaker 2.5 g of the sample, weighed to the nearest mg, in 5 ml of hot dilute hydrochloric acid TS. Cool, transfer to a 250-ml volumetric flask, dilute to volume with water, and mix. Transfer 50 ml of the solution to a 400-ml beaker, add 100 ml of water, 25 ml of sodium hydroxide TS, 40 mg of murexide indicator preparation and 3 ml of naphthol green TS. An alternative indicator is hydroxynaphthol blue, of which 0.25 g is used. In this case the naphthol green TS is omitted. Titrate with 0.05 M disodium ethylenediaminetetraacetate until the solution is

deep blue in colour. Each ml of 0.05 M disodium ethylenediaminetetraacetate is equivalent to 9.311 mg of $C_6H_{10}CaO_4$.

Calcium saccharin

Weigh accurately about 0.5 g of the sample and transfer quantitatively to a separator with the aid of 10 ml of water. Add 2 ml of dilute hydrochloric acid TS, and extract the precipitated saccharin first with 30 ml, then with five 20 ml portions, of a liquid composed of 9 volumes of chloroform and 1 volume of ethanol. Filter each extract through a small filter paper moistened with the solvent mixture. Evaporate the combined filtrates on a steam bath to dryness with the aid of a current of air. Dissolve the residue in 75 ml of hot water, cool, add phenolphthalein TS, and titrate with 0.1 N sodium hydroxide. Perform a blank determination, and make any necessary correction. Each ml of 0.1 N sodium hydroxide is equivalent to 20.22 mg of $C_{14}H_8CaN_2O_6S_2$.

Calcium sorbate

Weigh to the nearest mg, 0.25 g of the dried sample. Dissolve in 35 ml of glacial acetic acid and 4 ml of acetic anhydride in a 250-ml glass-stoppered flask, warming to effect solution. Cool to room temperature, add 2 drops of crystal violet TS and titrate with 0.1 N perchloric acid in glacial acetic acid to a blue-green end point which persists for at least 30 sec. Perform a blank determination and make any necessary correction. Each ml of 0.1 N perchloric acid is equivalent to 13.12 mg of $C_{12}H_{14}CaO_4$.

Carnauba wax

Weigh accurately about 5 g of the sample into a 250-ml flask, add a solution of 2 g of potassium hydroxide in 40 ml ethanol, and boil gently under reflux for 1 h or until saponification is complete. Transfer the content of the flask to a glass-stoppered extraction cylinder (approximately 30 cm in length, 3.5 cm in diameter and graduated at 40, 80 and 130 ml).Wash the flask with sufficient alcohol to achieve a volume of 40 ml in the cylinder, and complete the transfer with warm and then cold water until the total volume is 80 ml. Finally wash the flask with a few ml of petroleum ether, add the washings to the cylinder, cool the contents of the cylinder to room temperature and add 50 ml of petroleum ether. Insert the stopper and shake the cylinder vigorously for at least 1 min, and allow both layers to become clear. Siphon the upper ether layer as completely as possible without removing any of the lower layer, collecting the ether fraction in a 500-ml separator. Repeat extraction and siphoning at least six times with 50-ml portions of petroleum ether, shaking vigorously each time. Wash the combined extracts, with vigorous shaking, with 25-ml portions of 10% ethanol until the wash water is neutral to phenolphthalein, and discard the washings. Transfer the ether extract to a tared beaker and rinse the separator with 10-ml of ether, adding the rinsings to the beaker. Evaporate the ether at a steam bath just to dryness, and dry the residue to constant weight, preferably at 75° to 80° under vacuum of not more than 200 mm of Hg, or at 100° for 30 min. Cool in a desiccator and weigh to obtain weight of unsaponifiable matter. Dissolve the residue in 50 ml of warm neutral ethanol and titrate with 0.02N sodium hydroxide using phenolphthalein as indicator. Each ml of 0.02N sodium hydroxide is equivalent to 5.659 mg of fatty acids, calculated as oleic acid. Subtract the calculated weight of fatty acids from the weight of the residue to obtain the corrected weight of unsaponifiable matter in the sample.

Carbon-di-oxide	Transfer a 1 in 3 potassium hydroxide solution into a gas pipette of adequate volume. Measure accurately about 1,000 ml of the sample into a gas burette containing a 1 in 10 sodium chloride solution. Transfer the sample into the gas pipette and shake well.
Carthamus yellow	Transfer about 0.02 g of the sample, accurately weighed, in a 200-ml volumetric flask; dissolve in and dilute to volume with citric acid/disodium hydrogen phosphate buffer solution (pH 5.0), and centrifuge if necessary. Determine the absorbance (A) at an absorption maximum in 400-408 nm

in a 1-cm cell with the buffer solution as a blank.

Carthamus red	Transfer about 0.01 g of the sample, accurately weighed, in a 300-ml ground stoppered flask, add 150 ml of dimethylformamide (DMFA), dissolve by shaking occasionally and allow stand for 2 hours. Filter this solution through a glass filter into a 200-ml volumetric flask. Wash the flask and filter with two 25-ml portions of DMFA, combine the filtrate and the washings, add DMFA to volume and mix. Dilute if necessary. Determine the absorbance (A) at the maximum absorbance in the range of 525-535 nm using a 1-cm cell with DMFA as a blank.
Chlorophylins	Weigh accurately about 1 g of the sample, dried previously at 100° for 1 h, then dissolve in 20 ml Phosphate Buffer Solution (pH 7.5) and dilute to 1000 ml with distilled water. Dilute 10 ml of this solution to 100 ml with Phosphate Buffer solution (pH 7.5). Measure the optical density of the final solution (0.001% w/v) in a suitable spectrophotometer, using a 1 cm cell and slit width of 0.10 mm at 403-406 nm, recording the maximum within this range. The percentage of sodium copper chlorophyllin is given by the expression: 565 x weight of sample (g) optical density x 10. This formula was derived on the assumption that 100% pure sodium copper chlorophyllin has a specific absorbance of 565.
Chlorophylls, copper complexes	Accurately weigh about 100 mg of the sample and dissolve in diethyl ether, making the volume to 100 ml. Dilute 2 ml of this solution to 25 ml with diethyl ether. The concentration of the sample should not give an absorbance at 660.4 nm that is in excess of the working range for Absorbance measurements, i.e., not in excess of 0.7. Measure the absorbances (A) of the solution in a 1 cm cell against a diethyl ether blank at 667.2 nm, 654.4 nm, 649.8 nm and 628.2 nm. (The latter two wavelengths being the absorbance maxima in diethyl ether for copper phaeophytin a and copper phaeophytin b respectively). Calculate the concentration of the individual compounds in

micromoles per liter from the following equations:

Copper phaeophytin a = 45.6 A (649.8nm) - 2.75 A (628.2nm) + 3.10 A (667.2nm)- 35.4 A (654.4nm)

Copper phaeophytin b = -8.46 A (649.8nm)+ 20.7 A (628.2nm) - 1.69 A (667.2nm)+ 5.13 A (654.4nm)

Chlorophylls

Accurately weigh about 100 mg of the sample and dissolve in diethyl ether, making the volume to 100 ml. Dilute 2 ml of this solution to 25 ml with diethyl ether. The concentration of the sample should not give an absorbance at 660.4 nm that is in excess of the working range for Absorbance measurements, i.e., not in excess of 0.7. Measure the absorbances of the solution in a 1 cm cell against a diethyl ether blank at 660.4 nm, 642.0 nm, 667.2 nm and 654.4 nm. (These being the absorbance maxima in diethyl ether for chlorophyll a, chlorophyll b, phaeophytin a, and phaeophytin b, respectively). In addition measure at 649.8 nm and 628.2 nm. To the remaining diluted solution add one crystal of oxalic acid and after dissolution and mixing, remeasure the absorbances at the same wavelengths. "delta A" is the difference between the absorbances between the absorbance at the respective wavelengths, before and after addition of oxalic acid.

Cholic acid

Transfer about 400 mg of the dried sample, accurately weighed, into a 250-ml flask, add 20 ml of water and 40 ml of ethanol, cover with a watch glass, heat gently on a steam bath until dissolved and cool. Add 5 drops of phenolphthalein TS and titrate with 0.1 N sodium hydroxide to the first pink colour that persists for 15 sec. Perform a blank determination and make any necessary corrections. Each ml of 0.1 N sodium hydroxide is equivalent to 40.86 mg of $C_{24}H_{40}O_5$.

Citric acid

Weigh, to the nearest mg, 2.5 g of the sample and place in a tared flask. Dissolve in 40 ml of water and titrate with 1 N sodium hydroxide, using phenolphthalein TS as the indicator. Each ml of 1 N sodium hydroxide is equivalent to 64.04 mg of $C_6H_8O_7$.

Cyclohexyylsu lfamic acid

Transfer about 350 mg, accurately weighed, into a 250 ml flask. Dissolve the sample in 50 ml of water, add phenolphthalein TS, and titrate with 0.1N sodium hydroxide. Each ml of 0.1N sodium hydroxide is equivalent to 17.82 mg of $C_6H_{13}NO_3S$.

Curdlan

Transfer about 100 mg of the sample, accurately weighed, into a 100-ml volumetric flask and dissolve in about 90 ml of 0.1 N sodium hydroxide. Add 0.1 N sodium hydroxide to volume and mix well. Transfer 5 ml of the solution into a 100-ml volumetric flask, add water to volume and mix well. Quantitatively transfer 1 ml of the solution to a small flask or test tube, add 1 ml of a 5% (w/v) solution of reagent grade phenol in water and 5 ml of sulfuric acid TS. Shake vigorously and cool in ice-cold water. Prepare a blank and a reference standard solution in the same manner, using 0.1 ml of water and 100 mg of reagent grade glucose, respectively. Determine the absorbances of the sample solution and the reference standard solution in 1-cm cells at 490 nm with a suitable spectrophotometer, using the blank solution as the blank.

Curcumin

Accurately weigh about 0.08 g of the sample in a 200-ml volumetric flask and dissolve by shaking with ethanol. Make up to volume with ethanol and mix. Pipette 1.0 ml of solution into a 100-ml volumetric flask and make up to volume with ethanol. Determine the absorbance (A) at 425 nm in a 1-cm cell. Calculate the total colouring matters content of the sample.

Cupric sulfate

Weigh 1 g of the sample to the nearest 0.1 mg, and dissolve in 50 ml of water. Add 4 ml of acetic acid and 3 g of potassium iodide, and titrate the liberated iodine with 0.1 N sodium thiosulfate, using starch TS as the indicator. Perform a blank determination, and make any necessary correction. Each ml of 0.1 N sodium thiosulfate is equivalent to 24.97 mg of $CuSO_4 \cdot 5H_2O$.

Desoxycholic acid	Transfer about 500 mg of the dried sample, accurately weighed, into a 250-ml flask, add 20 ml of water and 40 ml of ethanol, cover with a watch glass, heat gently on a steam bath until dissolved and cool. Add 5 drops of phenolphthalein TS and titrate with 0.1 N sodium hydroxide, to the first pink colour that persists for 15 sec. Perform a blank determination and make any necessary corrections. Each ml of 0.1 N sodium hydroxide is equivalent to 39.26 mg of $C_{24}H_{40}O_4$.
Diammonium hydrogen phosphate	Dissolve about 600 mg of the sample, accurately weighed, in 40 ml of water and titrate to a pH of 8.0 with 0.1 N sulfuric acid. Each ml of 0.1 N sulfuric acid is equivalent to 13.21 mg of $(NH_4)_2HPO_4$.
Dicalcium pyrophosphate	Dissolve about 200 mg of the sample accurately weighed in 10 ml of dilute hydrochloric acid TS. Add about 120 ml of water and cool for 30 min, keeping the volume and pH of the solution constant during the cooling period by adding dilute hydrochloric acid or water if necessary. After cooling add 25 ml of 0.05 M disodium ethylenediamine tetraacetate and dilute to 200 ml with water. Neutralize with strong ammonia TS. Add 10 ml of buffer solution (pH 10) and a few drops of eriochrome black TS. Titrate with 0.05 M zinc sulfate. Each ml of 0.05 M disodium ethylenediamine tetraacetate is equivalent to 6.352 mg of Ca2P2O7.
Dichloromethane	The standard solution prepared above is diluted to a series of standards in the range of approximately 10 to 300 ppm (mg/kg) except for propylene oxide which is made in the range of 0.06 to 2.4 (w/w%). Standards and unknowns are injected into the gas chromatograph in the range of 1 to 5 µl (using split injection mode) and the peak areas determined by electronic integration. A standard curve is constructed from these dilutions by plotting peak area against concentration for

each analyte. The concentration of additives and by-products are determined by comparison to the standard curve. The sum of the concentrations of the impurities and stabilizers must be less than 1.0%.

Diethyl tartrate	Erlenmeyer flask containing a few boiling stones. Add to this flask, and, simultaneously, to a similar flask for a residual blank titration 25.0 ml of 0.5 N ethanolic potassium hydroxide. Connect each flask to a reflux condenser, and reflux the mixtures on a steam bath for exactly 1 h. Allow the mixtures to cool, add 10 drops of phenolphthalein TS to each flask, and titrate the excess alkali in each flask with 0.5 N hydrochloric acid.
Dimethyl dicarbonate	Introduce about 70 ml of pure acetone into a 150-ml glass beaker. Using a disposable 2 ml syringe weigh 1.0–1.3 g of the sample to an accuracy of ±0.1 mg into the glass beaker. Pipette exactly 20 ml dibutyl amine solution (add chlorobenzene to 120 g dibutyl amine until the 1 L mark is reached) while stirring. Titrate the solution potentiometrically with 1N hydrochloric acid. Run a blank test.
Diaryl thiodipropiona te	Weigh out 0.700 g of the sample, transfer to a 250-ml Erlenmeyer flask, add 100 ml of acetic acid and 50 ml of ethanol, and heat the mixture gently until the sample dissolves completely. Add 3 ml of hydrochloric acid and 4 drops of p-ethoxy-chrysoidin TS and immediately titrate with 0.1 N bromide-bromate TS. As the end-point is approached (pink colour), add 4 more drops of the indicator solution and continue the titration, dropwise, to a colour change from red to pale yellow. Perform a blank determination and make any necessary correction. Each ml of 0.1 N bromide-bromate TS is equivalent to 25.74 mg of $C_{30}H_{58}O_4S$. Convert to percentage and subtract thiodipropionic acid content determined in the Acidity test to obtain percentage of $C_{30}H_{58}O_4S$.

Dioctyl sodium sulfosuccinate	Sample solution: Transfer about 3.8 g of the sample, previously dried at 105° for 2 h and accurately weighed, into a 500-ml volumetric flask, dissolve in chloroform. Dilute to volume with the same solvent, and mix. Tetra-n-butylammonium iodide solution: Transfer 1.250 g of tetra-n-butylammonium iodide to a 500-ml volumetric flask, dilute to volume with water, and mix. Salt solution: Dissolve 100 g of anhydrous sodium sulfate and 10 g of sodium carbonate in sufficient water to make 1000 ml. Pipet 10.0 ml of the "Sample solution" into a 250-ml flask, and add 40 ml of chloroform, 50 ml of "Salt solution", and 10 drops of bromophenol blue TS. Titrate with "Tetra-n-butylammonium iodide solution" to the first appearance of a blue colour in the chloroform layer after vigorous shaking.
Dipotassium 5'-guanylate	Weigh accurately about 0.5 g of the sample, dissolve in and make to 1,000 ml with 0.01 N hydrochloric acid. Take 10 ml of this solution and dilute with 0.01 N hydrochloric acid to 250 ml. Determine the absorbance A of the solution in a 1-cm cell at the wave length of 260 nm using 0.01 N hydrochloric acid as the reference.
Dipotassium 5'-inosinate	Weigh accurately about 0.5 g of the sample, dissolve in and make to 1,000 ml with 0.01 N hydrochloric acid. Take 10 ml of this solution and dilute with 0.01 N hydrochloric acid to 250 ml. Determine the absorbance A of the solution in a 1-cm cell at the wave length of 250 nm using 0.01 N hydrochloric acid as the reference.
Dipotassium hydrogen phosphate	Into a 250-ml beaker transfer about 6.5 g of the dried sample, accurately weighed. Add 50 ml of 1 N hydrochloric acid and 50 ml of water, and stir until the sample is completely dissolved. Place the electrodes of a suitable pH meter in the solution and titrate the excess acid with 1N sodium hydroxide to the inflection point occurring at about pH 4. Record the buret reading and calculate

the volume (A) of 1N hydrochloric acid consumed by the sample. Continue the titration with 1N sodium hydroxide until the inflection point occurring at about pH 8.8 is reached, record the buret reading, and calculate the volume (B) of 1 N sodium hydroxide required in the titration between the two inflection points (pH 4 to pH 8.8).

Disodium ethylenediami netetraacetate	Transfer about 5 g, accurately weighed, of the sample, into a 250-ml volumetric flask, dissolve in water, dilute to volume and mix, to give the assay preparation. Place about 200 mg, accurately weighed, of reagent grade calcium carbonate of known purity in a 400-ml beaker, add 10 ml of water and swirl to form a slurry. Cover the beaker with a watch glass and introduce 2 ml of dilute hydrochloric acid TS from a pipette inserted between the lip of the beaker and the edge of the watch glass. Swirl the contents of the beaker to dissolve the calcium carbonate. Wash down the outer surface of the pipette, the watch glass and the sides of the beaker, and dilute to about 100 ml with water. While stirring the solution, preferably with a magnetic stirrer, add about 30 ml of the assay preparation from a 50-ml burette. Add 15 ml of sodium hydroxide TS, 300 mg of hydroxynaphthol blue indicator and continue the titration with the assay preparation to a blue end point.
Dodecyl gallate	Weigh accurately about 0.2 g of the dried sample into a 400-ml beaker. Add 150 ml of water and heat to boiling.
Dl-malic acid	Dissolve about 2 g of the sample, accurately weighed, in 40 ml of recently boiled and cooled water, add 2 drops of phenolphthalein TS and titrate with 1 N sodium hydroxide to the first appearance of a faint pink colour which persists for at least 30 sec. Each ml of 1 N sodium hydroxide is equivalent to 67.04 mg of $C_4H_6O_5$.

Erythorbic
acid

Weigh accurately about 0.4 g of the sample, previously dried, and dissolve in a mixture of 100 ml of water, recently boiled and cooled, and 25 ml of diluted sulfuric acid TS. Titrate the solution immediately with 0.1 N iodine, adding starch TS near the end point. Each ml of 0.1 N iodine is equivalent to 8.806 mg of C6H8O6.

Erythritol

Determine the erythritol content of the sample by liquid chromatography.

Mobile phase: Deionized water

Standard Solution: Transfer about 2 g of Standard Erythritol, previously dried in a vacuum desiccator at 70° for 6 hr and accurately weighed to the nearest 0.1 mg (W), into a 50 ml volumetric flask, dissolve in and dilute to volume with deionized water and mix. Filter the solution through a disposable 0.45 μm filter before use in the 'Procedure'.

Assay: Prepare as directed for 'Standard Solution', using about 2 g of the sample, previously dried in a vacuum desiccator at 70° for 6 h and accurately weighed to the nearest 0.1 mg (w).

Chromatographic System: Use a high-pressure liquid chromatography equipped with a constant-flow pulseless pump and fitted with a sensitive differential refractive index detector such as the RID-6A or equivalent. The column is packed with a strong cation exchange resin in the hydrogen form, such as MCI Gel-CK08EH, Shodex KC-811 or equivalent, consisting of a macroreticular sulfonated polystyrene-divinyl benzene copolymer, 8% cross linked, 9 μm particle size. The column temperature is 60°. . The sample injector is preferably of the fixed-loop type (manual or automatic), capable of accurately injecting 30 μl. The integrator can be any modern data acquisition system with recording and processing capabilities. The operating flow rate is about 0.5 ml/min. The maximum pressure of the total system is about 50 kgf/cm^2.

System Start-up: Connect the injector outlet to the column inlet, and connect the column outlet directly to waste. Activate the pump and elute the system at a flow rate of 0.1 ml/min. Set the pressure limit control to about 15 kgf/cm2 above the normal operating pressure. Increase the flow rate by increments of 0.1 ml/min up to the operating rate, and elute the column for 2 hours. Connect the column outlet to the detector tube, flush both the reference and sample cells for 30 min, and then zero the refractometer and adjust the sensitivity

System Suitability Test: The area responses of triplicate 30-µl injections of the Standard Solution show a relative standard deviation (100 × standard deviation/mean peak area) of not more than 1.0%).

Procedure: Chromatograph triplicate 30-µl portions of the Standard Solution and record the mean of the erythritol peak areas as A. In a similar manner, chromatograph triplicate 30-µl portions of the Assay Solution and record the mean of the erythritol peak areas as a. Calculate the percentage of erythritol in the sample by the formula: % Erythritol = 100(W/w)(a/A).

Ethyl alcohol

The specific gravity of the sample is not higher than 0.8096 at 25°/25° (equivalent to 0.8161 at 15.56°/15.56°).

Ethyl hydroxyethyl cellulose

Accurately weigh approximately 50 mg of the sample into a 100-ml round bottomed flask; add 10 ml of chromic acid solution and immerse the flask two thirds into an oil bath. The rest of the apparatus is fixed to the flask, and nitrogen is blown through at a rate of 1-2 bubbles per sec. The temperature of the bath is gradually raised over 30 min to 155° and held. Distillation starts at 135-140°. When 5 ml has been distilled, 5 ml of boiled distilled water is added from the graduated 50 ml dropping funnel. This procedure is continued until 50 ml of water has been added and consequently 55 ml of faintly yellow distillate has been collected. The distillate is quantitatively transferred to a flask and the distillate is titrated with 0.020N sodium hydroxide (carbon dioxide-free) to a phenolphthalein end-point. The solution is boiled 1 min and cooled to room temperature. The titration is continued until the pink colour remains stable for 10 sec. About 0.5 g of sodium hydrogen carbonate is added to the titrated solution followed by 10 ml of 10% sulfuric acid. When carbon dioxide evolution has ceased, 1 g of potassium iodide is added; the flask is shaken and kept 5 min in the dark. Liberated iodine is titrated with 0.020N sodium thiosulfate using 1% starch solution as the indicator. Chromium trioxide solution (10 ml) is distilled and titrated as described above to provide a blank test. It is necessary to run a new blank when a new chromium trioxide solution has been prepared or if changes have been made in the apparatus.

Ethyl p-hydroxybenzoate	Weigh, to the nearest mg, 2 g of the dried sample and transfer into a flask. Add 40 ml of N sodium hydroxide and rinse the sides of the flask with water. Cover with a watch glass, boil gently for 1 h and cool. Add 5 drops of bromothymol blue TS and titrate the excess sodium hydroxide with N sulfuric acid, comparing the colour with a buffer solution (pH 6.5) containing the same proportion of indicator. Perform a blank determination with the reagents and make any necessary correction. Each ml of N sodium hydroxide is equivalent to 166.18 mg of $C_9H_{10}O_3$.
Ethyl maltol	Standard solution
	Transfer about 50 mg of Ethyl Maltol Reference Standard (available from the United States Pharmacopoeia, 12601 Twinbrook Parkway, Rockville, MD 20852, USA), or equivalent, accurately weighed, into a 250-ml volumetric flask, dilute to volume with 0.1 N hydrochloric acid, and mix. Pipet 5 ml of this solution into a 100-ml volumetric flask, dilute to volume with 0.1 N

hydrochloric acid, and mix.

Sample solution

Transfer about 50 mg of the sample, accurately weighed, into a 250-ml volumetric flask, dilute to volume with 0.1 N hydrochloric acid, and mix. Pipet 5 ml of this solution into a 100-ml volumetric flask, dilute to volume with 0.1 N hydrochloric acid, and mix.

Procedure

Determine the absorbance of each solution in a 1-cm quartz cell at the absorption maximum (about 276 nm) using 0.1 N hydrochloric acid as the blank. Calculate the percent of Ethyl maltol in the sample by the formula: % of Ethyl maltol = $100 \times WS \times AA / (AS \times WA)$; where, AA is the absorbance of the sample solution; AS is the absorbance of the standard solution; WA is the weight in mg of the sample in the sample solution; WS is the weight in mg of the reference standard in the standard solution

Ethyl methyl ketone

Determine by gas-liquid chromatography (see Volume 4)(the GLC method allow the limit test for volatile organic impurities, such as alcohols, aldehydes, ketones; the total of these impurities shall not be more than 0.5%) using an instrument containing a thermal conductivity detector. Prepare a 4-m x 6-mm column consisting of a blend of equal quantities of 20% Carbowax 20 M on acid-washed, 60/80 mesh Chromosorb W, and 20% tetra-hydroxyethyl ethylenediamine on 30/60 mesh Chromosorb P, or use other suitable column materials capable of separating ethyl methyl ketone and the impurities whose retention times are listed below. Observe the following operating conditions during the determination: Sample size: 10 µl; Column temperature: about 80°; Helium flow rate: 30 to 32 ml per min; Detector voltage: 8.0 V; The approximate retention times, in min, are as follows: acetone - 7; ethyl acetate - 9; ethyl methyl ketone - 11; tertiary-butanol - 14; methanol - 15; ethanol - 19; 2-butanol - 30; propanol – 36

Fast red E	Proceed Titration with Titanous Chloride using the following: Weight of sample: 0.5-0.6 g; Buffer: 15 g sodium hydrogen tartrate; Weight: (D) of colouring matters equivalent to 1.00 ml of 0.1 N TiCl$_3$: 12.56 mg
Fast green fcf	Proceed as directed under Total Content by Titration with Titanous. Chloride (see Volume 4) using the following: Weight of sample: 1.9 - 2.0 g Buffer: 15 g sodium hydrogen tartrate Weight (D) of colouring matters equivalent to 1.00 ml 0.1 N TiCl3: 40.45 mg
Ferric ammonium citrate	Transfer about 1 g of the sample, accurately weighed, into a 250 ml glass-stoppered Erlenmeyer flask, and dissolve in 25 ml of water and 5 ml of hydrochloric acid. Add 4 g of potassium iodide, stopper, and allow to stand protected from light for 15 min. Add 100 ml of water, and titrate the liberated iodine with 0.1 N sodium thiosulfate, using starch TS as the indicator. Perform a blank determination and make any necessary correction. Each ml of 0.1 N sodium thiosulfate is equivalent to 5.585 mg of iron (Fe).
Ferrocyanides of calcium, potassium and sodium	Weigh 3 g of the sample to the nearest 0.1 mg and transfer into a 400-ml beaker. Dissolve in 225 ml of water, and add cautiously about 25 ml of sulfuric acid TS. Add, with stirring, 1 drop of orthophenanthroline TS, and titrate with 0.1 N ceric sulfate until the colour changes sharply from orange to pure yellow. Each ml of 0.1 N ceric sulfate is equivalent to 50.83 mg of $Ca_2Fe(CN)_6 \cdot 12H_2O$; 42.24 mg of $K_4Fe(CN)_6 \cdot 3H_2O$ or 48.41 mg of $Na_4Fe(CN)_6 \cdot 10H_2O$.
Ferrous ammonium	Weigh accurately about 0.3 g of the sample into a 250 ml conical flask, add 25 ml of dilute sulfuric acid (16% v/v) and dissolve with heating. Cool and add 75 ml of water. Add 0.1 ml of

phosphate

ferroin indicator solution (0.1% w/v in water). Titrate immediately with 0.1 N cerium sulfate until the colour changes from orange to light bluish-green. Each ml of 0.1 N cerium sulfate is equivalent to 5.585 mg of iron (II).

Ferrous gluconate

Dissolve about 1.5 g of the dried sample, accurately weighed, in a mixture of 75 ml of water and 15 ml of dilute sulfuric acid TS in a 300-ml Erlenmeyer flask, and add 250 mg of zinc dust. Close the flask with a stopper containing a Bunsen valve, and allow to stand at room temperature for 20 min. Then filter through a Gooch crucible containing a glass fibre filter paper coated with a thin layer of zinc dust, and wash the crucible and contents with 10 ml of dilute sulfuric acid TS, followed by 10 ml of water. Add orthophenanthroline TS and titrate the filtrate in the suction flask immediately with 0.1 N ceric sulfate. Perform a blank determination, and make any necessary correction. Each ml of 0.1 N ceric sulfate is equivalent to 44.61 mg of $C_{12}H_{22}FeO_{14}$.

Ferrous sulfate, dried

Dissolve about 1 g of the sample, accurately weighed, in a mixture of 25 ml of 2 N sulfuric acid and 25 ml of recently boiled and cooled water, add orthophenanthroline TS, and immediately titrate with 0.1 N ceric sulfate. Perform a blank determination, and make any necessary correction. Each ml of 0.1 N ceric sulfate is equivalent to 27.80 mg of $FeSO_4 \cdot 7H_2O$.

Ferrous sulfate

Dissolve about 1 g of the sample, accurately weighed, in a mixture of 25 ml of 2 N sulfuric acid and 25 ml of recently boiled and cooled water, add orthophenanthroline TS, and immediately titrate with 0.1 N ceric sulfate. Perform a blank determination, and make any necessary correction. Each ml of 0.1 N ceric sulfate is equivalent to 27.80 mg of $FeSO_4 \cdot 7H_2O$.

Ferrous lactate

Transfer about 2 g of the dried sample, accurately weighed, to a 100-ml volumetric flask, dilute to the mark with water and mix. Pipet 20 ml of the sample solution into a 100-ml conical flask. Add

5 ml of formic acid (85% v/v). Titrate the solution with 0.1 N potassium permanganate until it turns pink. Each ml of 0.1 N potassium permanganate is equivalent to 23.40 mg of $C_6H_{10}FeO_6$.

Ferrous glycinate (processed with citric acid)

Dissolve about 1 g, accurately weighed, of the dried sample in a mixture of 150 ml of water and 10 ml of sulfuric acid TS in a 300-ml flask. Add 1 drop of orthophenanthroline TS, and immediately titrate with 0.1 N ceric sulfate. Perform a blank determination in an identical manner, and perform any necessary correction; 1 ml of 0.1 N ceric sulfate is equivalent to 5.585 mg of ferrous ion.

Fumaric acid

Transfer about 1 g of the sample, accurately weighed, into a 250-ml Erlenmeyer flask, add 50 ml of methanol, and dissolve the sample by warming gently on a steam bath. Cool, add phenolphthalein TS, and titrate with 0.5 N sodium hydroxide to the first appearance of a pink colour that persists for at least 30 sec. Perform a blank determination and make any necessary correction. Each ml of 0.5 N sodium hydroxide is equivalent to 29.02 mg of $C_4H_4O_4$.

Gellan gum

Processed as directed in the test for Carbon Dioxide Determination by Decarboxylation in the General Methods, Volume 4, using about 1.2 g of the sample weighed accurately.

L-glutamic acid

Dissolve about 200 mg of the sample, previously dried and weighed accurately, in 6 ml of formic acid, and add 100 ml of glacial acetic acid. Titrate with 0.1 N perchloric acid determining the end-point potentiometrically. Run a blank determination in the same manner and correct for the blank. Each ml of 0.1 N perchloric acid is equivalent to 14.713 mg of $C_5H_9NO_4$.

Glycerol

Weigh accurately about 1 g of the sample and dissolve in water to make 100 ml. Add 100 ml of 0.3% potassium periodate solution to a 5 ml portion of the solution, shake thoroughly, and allow to stand for 1 h. Add 1 ml of propylene glycol, allow to stand for 10 min, and titrate with 0.05 N sodium hydroxide, using 3 drops of phenol red TS as the indicator, until a pink colour persists. Perform a blank test in the same manner as the sample. Each ml of 0.05 N sodium hydroxide is equivalent to 4.605 mg of $C_3H_8O_3$.

Glycerol diacetate

Transfer about 1 g of the sample, accurately weighed, into a suitable pressure bottle, add 25 ml of 1 N potassium hydroxide and 15 ml of isopropanol, stopper the bottle, and wrap securely in a canvas bag. Heat in a water bath maintained at 98±2o for 1 h, allowing the water in the bath to just cover the liquid in the bottle. Remove the bottle from the bath, cool in air to room temperature, then loosen the wrapper, uncap the bottle to release any pressure, and remove the wrapper. Add 6 to 8 drops of phenolphthalein TS and titrate the excess alkali with 0.5 N sulfuric acid just to the disappearance of the pink colour. Perform a blank determination. Each ml of 0.5 N sulfuric acid is equivalent to 44.04 mg of C7H12O5.

Grape skin extract

In the absence of an assay method, a measurement of colour intensity by the following method may be used. Prepare approximately 200 ml of pH 3.0 citric acid - dibasic sodium phosphate buffer solution: Mix 159 volumes of 2.1% citric acid solution and 41 volumes of 0.16% dibasic sodium phosphate solution, and adjust the pH to 3.0, using the citric acid solution or dibasic sodium phosphate solution. Weigh accurately an adequate amount of the sample so that the measured absorbance is between 0.2 and 0.7, and add pH 3.0 citric acid – dibasic sodium phosphate buffer solution to make up a 100-ml solution. Measure the absorbance A of this solution in a 1 cm cell at the wavelength of maximum absorption around 525 nm, using pH 3.0 citric acid – dibasic sodium phosphate buffer solution as the blank.

Colour value = weight of sample (g)

Green s
Proceed as directed under Total Content by Titration with Titanous Chloride (see Volume 4), using the following: Weight of sample: 1.4-1.5 g; Buffer: 15 g sodium hydrogen tartrate Weight (D) of colouring matters equivalent to 1.00 ml 0.1 N $TiCl_3$: 28.83 mg

5'-guanylic acid
Weigh accurately about 0.5 g of the sample, dissolve in and make to 1,000 ml with 0.01 N hydrochloric acid. Take 10 ml of this solution and dilute with 0.01 N hydrochloric acid to 250 ml. Determine the absorbance A of the solution in a 1-cm cell at the wave length of 260 nm using 0.01 N hydrochloric acid as the reference.

Helium
Determine by Gas chromatography using the following conditions: Column - material: stainless steel - length: 6 m; internal diameter: 4 mm; packing material: PoraPak Q, or equivalent Carrier; gas: Helium (99.99% (v/v); flow 40 ml/min; Detector: thermal conductivity detector; Injector: loop injector, Column temperature: 60^0; Detector temperature: 130^0

Procedure: Introduce a specimen of helium into a gas chromatograph by means of gas sampling valve. Select the operating conditions of the gas chromatograph so that the standard peak signal resulting from the following procedure corresponds to not less than 70% of the full-scale reading, and which permit complete separation of nitrogen and oxygen from helium, although nitrogen and oxygen may not be separated from each other. The peak response produced by the assay specimen exhibits a retention time corresponding to that produced by an air-helium certified standard (a mixture of 1.0% air in industrial-grade helium is available from most suppliers) and indicates not more than 1.0% air, by volume, when compared with the peak response of the air-helium certified standard, and not less than 99.0% of He, by volume.

Hexamethylen
Weigh, to the nearest 0.1 mg, 1 g of the sample, previously dried for 4 h over phosphorus

etetraamine	pentoxide. Transfer into a beaker and add 40.00 ml of N sulfuric acid. Boil gently adding water from time to time, if necessary, until the odour of formaldehyde is no longer perceptible. Cool, add 20 ml of water, and methyl red TS and titrate the excess acid with N sodium hydroxide. Each ml of N sulfuric acid is equivalent to 35.05 mg of C6H12N4.
4-hexylresorcino l	Indicator solution: Mix 1 g of soluble starch with 10 mg of mercuric iodide and sufficient cold water to make a thin paste. Add 200 ml of boiling water, and boil for 1 min with continuous stirring. Cool and use only the clear solution.

Procedure: Accurately weigh into a 250-ml iodine flask 70-100 mg of the sample, previously dried for 4 hours at room temperature over silicagel, and dissolve the sample in 10 ml of methanol. Add 30.0 ml of bromide/bromate TS, then quickly 5 ml of hydrochloric acid and insert the stopper in the flask immediately. Cool the flask under running water to room temperature, shake vigorously for 5 min, then set aside for 5 min. Add 6 ml of potassium iodide TS around the stopper, cautiously loosen the stopper, again insert the stopper tightly, and swirl gently. Add 1 ml of chloroform, and titrate the liberated iodine with 0.1N sodium thiosulfate, adding 3 ml of Indicator solution as the end point is approached. Perform a blank determination. |
| Hydrochloric acid | Tare accurately a 125-ml glass-stoppered conical flask containing 50 ml of 1N sodium hydroxide. Partially fill, without the use of vacuum, a 10-ml serological pipet from near the bottom of a representative sample, remove any acid adhering to the outside and discard the first ml flowing from the pipet. Hold the tip of the pipet just above the surface of the sodium hydroxide solution, and transfer between 2.5 and 3 ml of the sample into the flask, mix the contents, and weigh |

accurately to obtain the weight of the sample. Add methyl orange TS and titrate the excess of sodium hydroxide with 1N hydrochloric acid. Each ml of 1N sodium hydroxide is equivalent to 36.46 mg of HCl.

Hydrogen peroxide

Accurately weigh a volume of the sample equivalent to about 300 mg of H2O2 into a 100-ml volumetric flask, dilute to volume with water, and mix thoroughly. To a 20-ml portion of this solution add 25 ml of diluted sulfuric acid TS, and titrate with 0.1 N potassium permanganate. Each ml of 0.1 N potassium permanganate is equivalent to 1.701 mg of H2O2.

1-hydroxyethyli dene-1,1-diphosphonic acid

Place about 3 g of the sample, accurately weighed (w) into a beaker and add 100-150 ml of water. Stir the solution with a magnetic stirrer (maintain throughout titration). Insert pH electrode(s) and record the pH value. Titrate with 1 mol/l sodium hydroxide and record pH (or millivolts) after every 1ml added. Stop the titration at pH 10. Plot the pH as a function of added sodium hydroxide and manually draw the titration curve. Two inflection points will be observed at around pH 3 and pH 8. Take only into account the inflection point at around pH 8. Trace the tangent to this inflection point in order to determine the end-point.

5'-inosinic acid

Weigh accurately about 0.5 g of the sample, dissolve in and make to 1,000 ml with 0.01 N hydrochloric acid. Take 10 ml of this solution and dilute with 0.01 N hydrochloric acid to 250 ml. Determine the absorbance A of the solution in a 1-cm cell at the wave length of 250 nm using 0.01 N hydrochloric acid as the reference.

Insoluble polyvinylpyrrolidone

Determine the nitrogen content according to Kjeldahl with a modified digestion. Place about 0.1 g, accurately weighed, in the digestion flask of the apparatus. Add 1 g of a powdered mixture of 10 parts of potassium sulfate and 1 part of cupric sulfate, and wash down any adhering material from the neck of the flask with a fine jet of water. Add 7 ml of sulfuric acid, rinsed down the wall of the flask, then, while swirling the flask, add 1 ml of 30% hydrogen peroxide cautiously down the side of the flask (Do not add hydrogen peroxide during the digestion). Repeat the addition of 1 ml of 30% hydrogen peroxide (usually 3 to 6 times) until a clear, light green solution is obtained on heating the mixture. Heat for an additional 4 h. Cautiously add to the digestion mixture 20 ml of water and proceed to the steam distillation as directed under Nitrogen Determination (Kjeldahl Method).

Iron oxides

Weigh accurately about 0.2 g of the sample, add 10 ml of 5 N hydrochloric acid, and heat cautiously to boiling in a 200-ml conical flask until the sample has dissolved. Allow to cool, add 6 to 7 drops of 30% hydrogen peroxide solution and again heat cautiously to boiling until all the excess hydrogen peroxide has decomposed (about 2-3 min). Allow to cool, add 30 ml of water and about 2 g of potassium iodide and allow to stand for 5 min. Add 30 ml of water and titrate with 0.1 N sodium thiosulfate adding starch TS as the indicator towards the end of the titration. Each ml of 0.1N sodium thiosulfate is equivalent to 5.585 mg of Fe (III).

Isobutanol

Determine by gas-liquid chromatography (see Volume 4), using the following conditions: Column - length: 2.4 m; diameter: 6 mm; material: copper; packing: 23% Carbowax 1500; support: Chromosorb W (60/80 mesh); Carrier gas: Helium; Flow rate: 150 ml/min; Detector type: FID; Temperatures; injection port: 150^0; column: 70^0; detector: 150^0; Inject 1 to 5 µl of sample, obtain chromatogram and determine content of each constituent by the method of area normalization.

Isopropyl acetate

Transfer 25.0 ml of 1 N potassium hydroxide TS into a suitable heat-resistant pressure bottle provided with a tight closure that can be securely fastened, and then add 10 ml of isopropanol and a few pieces of glass rod. To the mixture in the pressure bottle add about 1.3 g of the sample contained in a sealed glass ampoule and accurately weighed. Cap the bottle, shake it vigorously to break the ampoule, and allow it to stand at room temperature for 30 min. Uncap the bottle, add phenolphthalein TS, and titrate with 0.5 N sulfuric acid to the disappearance of the pink colour. Perform a residual blank titration. Each ml of 0.5 N sulfuric acid is equivalent to 51.07 mg of $C_5H_{10}O_2$.

Konjac flour

The remainder, after subtracting from 100% the sum of the percentages of total ash, loss on drying and protein, represents the percentage of carbohydrate (glucomannans) in the sample.

Lactic acid

Weigh accurately a portion of the sample equivalent to about 3 g of lactic acid, transfer to a 250-ml flask, add 50 ml of 1N sodium hydroxide, mix, and boil for 20 min. Add phenolphthalein TS, titrate the excess alkali in the hot solution with 1N sulfuric acid, and perform a blank determination. Each ml of 1N sodium hydroxide is equivalent to 90.08 mg of $C_3H_6O_3$.

Lecithin

Purification of phosphatides: Wash about 10 g of the sample 3 times well with each 100 ml of acetone. The insoluble residue (phosphatides) obtained from assays carried out previously can also be used. Dissolve 5 g of these phosphatides in 10 ml of petroleum ether, and add 25 ml of acetone to the solution. Transfer approximately equal portions of the precipitate to each of two 40-ml centrifuge tubes using additional portions of acetone to facilitate the transfer. Stir thoroughly, dilute to 40 ml with acetone, stir again, chill for 15 min in an ice bath, stir again, and then centrifuge for 5 min. Decant the acetone, stir, chill, centrifuge, and decant as before. The solids after the second centrifugation require no further purification and may be used for preparing the phosphatide-acetone solution. To saturate about 16 litres of acetone, 5 g of the purified phosphatides are required.

Lecithin partially hydrolyzed

Purification of phosphatides: Wash about 10 g of the sample 3 times well with each 100 ml of acetone. The insoluble residue (phosphatides) is used. Residues (phosphatides) obtained from assays carried out previously can also be used. Dissolve 5 g of these phosphatides in 10 ml of petroleum ether, and add 25 ml of acetone to the solution. Transfer approximately equal portions of the precipitate to each of two 40-ml centrifuge tubes using additional portions of acetone to facilitate the transfer. Stir thoroughly, dilute to 40 ml with acetone, stir again, chill for 15 min in an ice bath, stir again, and then centrifuge for 5 min. Decant the acetone, stir, chill, centrifuge, and decant as before. The solids after the second centrifugation require no further purification and may be used for preparing the phosphatide-acetone solution. To saturate about 16 litres of acetone, 5 g of the purified phosphatides are required. Phosphatide acetone solution Add a quantity of purified phosphatides to sufficient acetone, previously cooled to a temperature of about 5o, to form a saturated solution, and maintain the mixture at this temperature for 2 h., shaking it vigorously at 15-min. intervals. Decant the solution through a rapid filter paper, a voiding the transfer of any undissolved solids to the paper and conducting the filtration under refrigerated conditions (not above 5°).

Procedure: If lecithin is plastic or semisolid, soften a portion of the sample by warming it in a water bath at a temperature not exceeding 60o and then mixing it thoroughly. Transfer about 2 g of a well-mixed sample, accurately weighed, into a previously tared 40-ml centrifuge tube, containing a glass stirring rod, and add 15 ml of Phosphatide-Acetone Solution from a buret. Warm the mixture in a water bath until the lecithin melts, but avoid evaporation of the acetone. Stir until the sample is completely disintegrated and dispersed, and then transfer the tube into an ice bath, chill for 5 min, remove from the ice bath, and add about one half of the required volume of Phosphatide- Acetone Solution, previously chilled for 5 min in an ice bath. Stir the mixture to complete dispersion of the sample, dilute to 40 ml with chilled Phosphatide-Acetone Solution (5o), again stir, and return the tube and contents to the ice bath for 15 min. At the end of the 15-min chilling period, stir again while still in the ice bath, remove the stirring rod, temporarily supporting it in a vertical upside-down position, and centrifuge the mixture immediately at about 2000 rpm for 5 min. Decant the supernatant liquid from the centrifuge tube, crush the centrifuged solids with the same stirring rod previously used, and refill the tube to the 40-ml mark with chilled (5°) Phosphatide-Acetone Solution and repeat the chilling, stirring, centrifugation, and decantation procedure previously followed. After the second centrifugation and decantation of the supernatant acetone, again crush the solids with the assigned stirring rod, and place the tube and its contents in a horizontal position at room temperature until the excess acetone has evaporated. Mix the residue again, dry the centrifuge tube and its contents at 105o for 45 min in a forced-draft oven, cool, and weigh.

Lacitol

Determine lactitol as well as other polyols resulting as by-products during the manufacture of lactitol by liquid chromatography. Principal by-product polyols are the hexitols: sorbitol, mannitol, galactitol (dulcitol), and lower polyols such as glycitols.

Apparatus: Liquid chromatograph with elevated temperature capability, differential refractometric detector and 0.45 μm membrane filter before column.

Column: Aminex HPX 87 (calcium form) with dimensions 300 x 7.8 mm, or equivalent column designed for carbohydrate analyses

Standards: Lactitol, sorbitol, mannitol; Eluent; Water (degassed).

Procedure: Equilibrate chromatography column to 85°. Adjust eluent flow rate through column to 0.6 ml/min. Accurately prepare an aqueous solution of sample about 40% by weight. Inject 10 μl of the 40% sample solution onto the column. Record the chromatogram for peaks occurring at the retention time of lactitol and thereafter. Approximate retention times for lactitol and other polyols using the recommended column are: Lactitol 12 min; Ribitol 15 min; Erythritol 16 min; Mannitol 18 min; Galactitol 20 min; Sorbitol 21 min; For Assay, compare the sample response relative to the response of a standard sample of lactitol of known purity. For other polyols, measure the area of all peaks occurring between Lactitol and Sorbitol. The sum of the areas of these peaks is not greater than 2.5 % of the dry weight of the sample.

Lithol rubine bk

Place about 0.2 g of the sample, accurately weighed, in a 500-ml Erlenmeyer flask and add 5 ml of sulfuric acid. Mix well and add 100 ml of ethanol. Shake well, heat on a water bath. Add a solution made by dissolving 20 g of sodium hydrogen tartrate in 100 ml of boiling water and mix with 20 ml of 30% sodium hydroxide solution, shaking vigorously. Titrate with 0.1 N titanous chloride. Each ml of 0.1 N titanous chloride is equivalent to 10.61 mg of $C_{18}H_{12}CaN_2O_6S$.

Lycopene extract from tomato

Total lycopene and Total carotenoids; Total lycopenes are determined by HPLC. Total carotenoids are determined spectrophotometrically.

HPLC system with a UV/VIS detector or a diode array detector, auto sampler or injector; Detector: 472 nm; Column: Select B (RP-C8) (250 x 4.6 mm, 5 µm) Merck no. 50984 or equivalent; Mobile phase: acetonitrile:methanol:dichloromethane:nhexane: N-ethyl-diisopropylamine 850:100:25:25:0.5 (v/v/v/v). Mix well and sonicate for 3-4 min in an ultrasonic bath; Flow rate: 0.7 ml/min; Injection volume: 10 µl; Run time: 12 min; Diluent solution Transfer 0.5 g BHT, 600 ml acetonitrile, 100 ml methanol, 150 ml dichloromethane and 150 ml n-hexane into a 1000-ml bottle. Mix well and sonicate for 3-4 min in an ultrasonic bath. Lycopene standard stock solution (500 mg/l) Weigh accurately (to ± 0.1 mg) about 50 mg all-*trans*-lycopene standard into a 100-ml volumetric flask and add 100 mg of α- tocopherol and 100 mg of BHT. Add toluene to volume and sonicate 1-2 min, mix well. Dispense to 8-ml amber vials. The solution is stable for six months when stored at -18°. Lycopene standard solutions Take one vial of the Lycopene standard stock solution and warm to 50° in a water bath for several minutes, shaking the solution occasionally to ensure that the lycopene particles are completely dissolved. Transfer 3 ml of this solution to a 25-ml amber volumetric flask and add the Diluent solution to volume and mix (Solution A). Take another vial of the Lycopene standard stock solution and treat as above. Transfer 4 ml of this solution to a second 25-ml amber flask and add the Diluent solution to volume and mix (Solution B). Solutions A and B are stable for at least 3 weeks if held at -18°. Prior to each use, determine spectrophotometrically the lycopene concentration in each solution (See Standardization of the Lycopene standard solutions). BHT solution (5000 mg/l) Weigh 2.5 g

BHT into a 500-ml storage bottle and add 500 ml dichloromethane. Keep the solution protected from light. This solution is stable for 3 months.

Sample solutions: Introduce a representative sample of the tomato extract into a vial and close it. Place the vial in a water bath at 50° for 30 minutes. (NOTE: The temperature should not exceed 60°). Stir the solution using a glass rod. Weigh accurately (to ±0.1 mg) 1.0 to 1.2 g of the sample into each of three 100-ml (VA) volumetric flasks (samples 1, 2 and 3) and add 10 ml of BHT solution and 40 ml of dichloromethane to each flask. Homogenize the solutions using an ultrasonic bath, cool the solutions to room temperature and bring each to volume with dichloromethane and mix (Solutions C). Transfer 5 ml (VB) of each Solution C to separate amber 50-ml (VC) volumetric flasks. Bring each to volume with the Diluent solution and mix well (Solutions D).

Procedure: Transfer 2.0 ml (VD) of each of Solutions A and B into 100-ml (VE) volumetric flasks and add 10 ml of ethanol and 10 ml of BHT solution. Bring the two solutions to volume with petroleum ether (Solutions E and F). Using a suitable UV/VIS spectrophotometer and 1-cm cell, determine the absorbances of these solutions at 472 nm using petroleum ether as a blank. Inject Solutions A and B into the chromatograph. Record the peak areas. Inject the three sample solutions (Solutions D) and record the peak areas of lycopene (the retention time of all isomers of lycopene is approximately 5 to 7 min and that for β-carotene is 8 to 9 min).The peak area of lycopene for the sample solutions should be between 80 and 120% of the standards, otherwise dilute the Solution C with the Diluent solution to bring the lycopene concentration to the desired range or increase the sample weight.

Total carotenoids: Using a volumetric pipette, transfer 2 ml (VF) of Solution D (see above) to an amber 100-ml (VG) volumetric flask. Add 10 ml of ethanol, bring to volume with petroleum ether and mix well. This is sample Solution G. Using a suitable UV/VIS spectrophotometer and 1-cm sample cells with covers, scan the spectrum of Solution G from 550 to 300 nm, using petroleum ether as the reference blank and measure the absorbance at the absorbance maximum (approximately 472 nm). The absorbance should be between 0.2 and 0.8.

Lycopene from blakeslea trispora

The HPLC method of assay is suitable for determination of total lycopenes (all-*trans*-lycopene and *cis*-lycopene isomers), all-*trans*lycopene, and other carotenoids. (Note: the predominant *cis* isomer detected in lycopene from *B. trispora* is 13-*cis*-lycopene.) Reagents (Note: all solvents should be HPLC-grade) Acetonitrile Methanol Acetone Hexane Methylene chloride Lycopene standard (purity 95% or higher; available from Vitatene S.A.)

Apparatus: VIS or UV/VIS spectrophotometer with a 1-cm light path optical cell HPLC system with either a VIS or UV/VIS detector or a suitable diode array detector, injector, column oven, and integrator Column: Vydac 218 TP54 5 m (4.6x250 mm) or equivalent HPLC conditions: Mobile phase: acetonitrile/methanol (40:60); Flow rate: 1 ml/min; Detection: 470 nm; Injection volume: 10 µl; Column temperature: 30°; Injector temperature: 10°; Run time: 15 min;

Standard solution: Weigh accurately about 25 mg lycopene standard into a 100-ml volumetric flask. Dissolve in 10 ml of methylene chloride and add hexane to volume. Pipet 1 ml of the above solution into a 50-ml volumetric flask and add acetone to volume.

Sample solution: Prepare as the standard solution: HPLC analysis Chromatograph the standard solution. The retention time of all-*trans*lycopene is approximately 11.5 to 12.5 min. The relative retention time of 13-*cis*-lycopene with respect to all-*trans*-lycopene is 1.25. The relative retention times for other carotenoids with respect to all-*trans*lycopene are 1.2 for β-carotene and 1.1 for γ-carotene. Record the total peak area of all-*trans*-lycopene and *cis*-lycopene isomers and calculate the response factor (RF) for lycopene as follows:

Wst x Pst RF = Ast x 5000; where, RF is the response factor for lycopene (AU ml/mg);

Ast is the total lycopene (all-*trans*-lycopene + *cis*-lycopene isomers) peak area; 5000 is the volume of the volumetric flask in which the standard was dissolved (100 ml) multiplied by dilution (50); Wst is the weight of the standard (mg); and Pst is the purity of the standard expressed as a proportion of lycopene in the lycopene standard (determined as described under Standard purity determination). Chromatograph the sample solution and record the following peak areas: A1 – all-*trans* lycopene; A2 – total lycopene (all-*trans*-lycopene + *cis*-lycopene isomers); A3 – other carotenoids; A4 – all carotenoids (all-*trans*-lycopene + *cis*-lycopene isomers + other carotenoids).

Magnesium chloride	Dissolve about 450 mg of the sample, accurately weighed, in 25 ml of water, add 5 ml of ammonia/ammonium chloride buffer TS and 0.1 ml of eriochrome black TS and titrate with 0.05 M disodium ethylenediaminetetraacetate until the solution is blue in colour. Each ml of 0.05 M disodium ethylenediaminetetra-acetate is equivalent to 10.16 mg of $MgCl_2 \cdot 6H_2O$.
Magnesium carbonate	Weigh 1 g of the sample to the nearest 0.1 mg, and transfer to a 250 ml conical flask. Pipette into the flask 50 ml of N sulfuric acid and swirl to dissolve. Titrate the excess acid with N sodium hydroxide solution, using methyl orange TS as indicator. Subtract from the volume of N sulfuric

acid consumed the number of ml of N sulfuric acid corresponding to the weight of Ca in the sample taken, using as a factor 20.04 mg of Ca for each ml of N sulfuric acid. The difference is the volume of N sulfuric acid used to neutralize the magnesium carbonate and each ml is equivalent to 12.16 of Mg.

Magnesium di-1-glutamate

Dissolve about 250 mg of the sample, previously dried and weighed accurately, in 6 ml of formic acid, and add 100 ml of glacial acetic acid. Titrate with 0.1 N perchloric acid determining the end-point potentiometrically. Run a blank determination in the same manner and correct for the blank. Each ml of 0.1 N perchloric acid is equivalent to 7.914 mg of $C_{10}H_{16}MgN_2O_8$. Calculate the content on the anhydrous basis.

Magnesium dl-lactate

Dissolve about 0.5 g of the dried sample, accurately weighed, in 25 ml of water, add 5 ml of ammonia/ammonium chloride buffer TS and 0.1 ml of ethylenediaminetetraacetate until the solution is blue in colour. Each ml of 0.05 M disodium ethylenediaminetetraacetate is equivalent to 10.12 mg of $Mg(C3H5O3)2$.

Magnesium hydrogen phosphate

Weigh accurately about 500 mg of the residue obtained in the test for Loss on Ignition, and dissolve it by heating in a mixture of 50 ml of water and 2 ml of hydrochloric acid. Cool, dilute to 100.0 ml with water, and mix. Transfer 50.0 ml of this solution into a 400-ml beaker, add 100 ml of water, and heat to 55 $^\circ$ to 60 $^\circ$. From a buret add 15 ml of 0.1 M disodium EDTA, add a magnetic stirring bar, and adjust with sodium hydroxide TS to pH 10 while stirring. Add 10 ml of ammonia-ammonium chloride buffer TS and 12 drops of eriochrome black TS, and continue the titration with 0.1 M disodium EDTA until the wine-red colour changes to pure blue. Calculate the weight, in mg, of $Mg_2P_2O_7$ in the residue.

Magnesium hydroxide	Transfer about 400 mg of the sample, previously dried at 105 ° for 2 h and accurately weighed, into a conical flask. Add 25 ml of 1 N sulfuric acid and, after solution is complete, add methyl red TS and titrate the excess acid with 1 N sodium hydroxide. Subtract from the volume of 1 N sulfuric acid consumed in the assay the volume of 1 N sulfuric acid corresponding to the weight of CaO in the sample taken for the assay using as a factor 28.04 mg of CaO for each ml of 1 N sulfuric acid. Each ml of 1 N sulfuric acid used to neutralize the magnesium hydroxide is equivalent to 29.16 mg of $Mg(OH)_2$.
Magnesium hydroxide carbonate	Dissolve about 1 g of the sample, accurately weighed, in 30 ml of 1 N sulfuric acid, add methyl orange TS, and titrate the excess acid with 1 N sodium hydroxide. From the volume of 1 N sulfuric acid consumed, deduct the volume of 1 N sulfuric acid corresponding to the content of calcium oxide in the weight of the sample taken for the assay. The difference is the volume of 1 N sulfuric acid equivalent to the magnesium oxide present. Each ml of 1 N sulfuric acid is equivalent to 20.15 mg of MgO.
Magnesium l-lactate	Dissolve about 0.5 g of the dried sample, accurately weighed, in 25 ml of water, add 5 ml of ammonia/ammonium chloride buffer TS and 0.1 ml of ethylenediaminetetraacetate until the solution is blue in colour. Each ml of 0.05 M disodium ethylenediaminetetraacetate is equivalent to 10.12 mg of $Mg(C_3H_5O_3)_2$.
Magnesium oxide	Ignite about 400 mg of the sample to constant weight at 800o to 825o in a tared platinum crucible. Weigh the residue accurately, dissolve in 25.0 ml of N sulfuric acid, boil gently to remove any carbon dioxide and cool. Add methyl red TS and titrate the excess acid with N sodium hydroxide. Subtract from the volume of N sulfuric acid consumed the number of ml of N sulfuric acid

corresponding to the weight of CaO in the sample taken, using as a factor 28.04 mg of CaO for each ml of N sulfuric acid. The difference is the volume of N sulfuric acid used to neutralize the magnesium oxide and each ml is equivalent to 20.16 mg MgO.

Magnesium silicate (synthetic)	Magnesium oxide: Weigh 1.5 g of the sample, to the nearest 0.1 mg, transfer into a 250 ml conical flask, add 50 ml of 1 N sulfuric acid, and digest on a steam bath for 1 h. Cool to room temperature, add methyl orange TS, and titrate the excess acid with 1 N sodium hydroxide. Each ml of 1 N sulfuric acid is equivalent to 20.15 mg of MgO.
	Silicon dioxide: Transfer about 0.7 g of the sample, weighed to the nearest 0.1 mg (W1), into a 150 ml beaker add 20 ml of 1 N sulfuric acid, and heat on a steam bath for 1.5 h. Decant the supernatant liquid through an ashless filter paper, and wash the residue, by decantation, three times with hot water. Treat the residue with 25 ml of water and digest on a steam bath for 15 min. Finally, transfer the residue to the filter paper and wash thoroughly with hot water. Transfer the filter paper and its contents to a platinum crucible. Heat to dryness, incinerate, then ignite strongly for 30 min, cool and weigh. Moisten the residue with water, and add 6 ml of concentrated hydrofluoric acid (warning: toxic, corrosive, must not contact skin; work with fume hood) and 3 drops of sulfuric acid TS. Evaporate to dryness, ignite for 5 min, cool and weigh.
	The loss in weight represents the weight of SiO2 (W3): %SiO2 = 100* (W2 - W3)/(W1); Where: W1 = Weight of sample taken; W2 = Weight after sulfuric acid treatment; W3 = Weight after concentrated hydrofluoric and sulfuric acid treatment
Magnesium sulfate	Accurately weigh about 0.5 g of the ignited sample, dissolve in 5 ml of hydrochloric acid TS, Dilute, dilute with water to 100 ml, and mix. Transfer 50 ml of this solution into a 250-ml conical

flask; add 10 ml of Ammonia/Ammonium Chloride Buffer TS and 0.1 ml of Eriochrome Black TS. Titrate with 0.05 M disodium EDTA until the colour of redpurple solution changes to blue. Each ml of 0.05 M disodium EDTA is equivalent to 12.04 mg of $MgSO_4$.

Maltol

Standard solution Transfer about 50 mg of Maltol Reference Standard (available from the United States Pharmacopoeia, 12601 Twinbrook Parkway, Rockville, MD 20852, USA), or equivalent, accurately weighed, into a 250-ml volumetric flask, dilute to volume with 0.1 N hydrochloric acid, and mix. Pipet 5 ml of this solution into a 100-ml volumetric flask, dilute to volume with 0.1 N hydrochloric acid, and mix. Assay solution. Transfer about 50 mg of the sample, accurately weighed, into a 250-ml volumetric flask, dilute to volume with 0.1 N hydrochloric acid, and mix. Pipet 5 ml of this solution into a 100-ml volumetric flask, dilute to volume with 0.1 N hydrochloric acid, and mix. Procedure Determine the absorbance of each solution in a 1-cm quartz cell at the absorption maximum (about 274 nm) using 0.1 N hydrochloric acid as the blank. Calculate the percent of Maltol

Methanol

Using the procedures for Gas chromatography, establish the following conditions: Column-length: 1.8 m; diameter: 4 mm; packing: 120-150 mesh Porapak R, or equivalent; Carrier gas: Nitrogen; Flow rate: 25 ml/min; Detector: FID; Temperatures; - injection port: 200^0; column: 160^0; detector: 210^0

Prepare a standard solution of 0.4% (v/v) methanol in dioxane. Adjust column temperature and/or gas flow rate so that methanol retention time is about 5-7 min. Adjust detector so that 8 µl of standard solution provides at least one-half scale deflection. Inject 5-10 µl sample, obtain chromatogram and determine methanol content by the method of area normalization.

Methyl p-hydroxybenzoate

Weigh, to the nearest mg, 2 g of the dried sample and transfer into a flask. Add 40 ml of N sodium hydroxide and rinse the sides of the flask with water. Cover with a watch glass, boil gently for 1 h and cool. Add 5 drops of bromothymol blue TS and titrate the excess sodium hydroxide with N sulfuric acid, comparing the colour with a buffer solution TS (pH 6.5) containing the same proportion of indicator. Perform a blank determination with the reagents and make any necessary correction. Each ml of N sodium hydroxide is equivalent to 152.2 mg of $C_8H_8O_3$

Microcrystalline cellulose

Transfer about 125 mg of the sample, accurately weighed, to a 300 ml Erlenmeyer flask, using about 25 ml of water. Add 50.0 ml of 0.5N potassium dichromate and mix. Carefully add 100 ml of sulfuric acid and heat to boiling. Remove from heat, allow to stand at room temperature for 15 min and cool in a water bath. Transfer the contents into a 250 ml volumetric flask, rinse flask with distilled water, add rinsings to the volumetric flask and dilute with water almost to volume. Allow the volumetric flask to reach room temperature (25°); then make up to volume with water and mix. Titrate a 50.0 ml aliquot with 0.1N ferrous ammonium sulfate using 2 or 3 drops of orthophenanthroline TS as the indicator and record the volume required as S in ml. Perform a blank determination and record the volume of 0.1N ferrous ammonium sulfate required as B in ml. Calculate the percentage of cellulose in the sample.

Nisin

Determination of nisin activity

Preparation of the test organism: *Lactococcus lactis subsp. cremoris* (ATCC 14365, NCDO 495) is subcultured daily in sterile separated milk by transferring one loopful to a McCartney bottle of litmus milk and incubating at 30°. Prepare inoculated milk for the assay by inoculating a suitable quantity of sterile skim milk with 2 percent of a 24 h culture, and place it in a water-bath at 30° for

90 min. Use immediately.

Standard stock solution: Dissolve an accurately weighed quantity of standard nisin in 0.02 N hydrochloric acid to give a solution containing 5,000 units/ml. Immediately before use, dilute the solution further with 0.02 N hydrochloric acid to give 50 units/ml. (NOTE: Nisin containing 2.5% nisin, minimum potency of 106 IU/g, obtainable from Sigma, St Louis, USA or Fluka, Buchs, Switzerland, may be used for the Standard stock solution, as well as the preparation under the name of Nisaplin, available from Danisco, Copenhagen, Denmark).

Sample solution: Weigh an amount of sample sufficient to ensure that corresponding tubes of the sample and standard series match, i.e., within close limits, the nisin content in the sample and standard is the same. Dilute the sample solution in 0.02 N hydrochloric acid to give an approximate concentration of 50 units of nisin per ml.

Resazurin solution: Prepare a 0.0125% w/v solution of resazurin in water immediately before use.

Procedure: Pipet graded volumes (0.60, 0.55, 0.50, 0.45, 0.41, 0.38, 0.34, 0.31, 0.28, 0.26 ml) of the 50 unit per ml sample and standard solutions into rows of 10 dry 6-inches x 5/8-inch bacteriological test tubes. Add 4.6 ml of the inoculated milk to each by means of an automatic pipetting device. (NOTE: The addition of inoculated milk is made in turn across each row of tubes containing the same nominal concentration, not along each row of ten tubes.) Place the tubes in a water bath at 30° for 15 min, then cool in an ice-water bath while adding 1 ml resazurin solution to each. Make the addition in the same order as for the addition of inoculated milk, using an automatic pipetting device. Thoroughly mix the contents of the tubes by shaking. Continue incubation at 30° in a water bath for a further 3-5 min. Examine the tubes under fluorescent light in a black matt-finish cabinet. Compare the sample tube of the highest concentration that shows the first clear difference in colour (i.e., has changed from blue to mauve) with tubes of the standard row of tubes to find the nearest match in colour. Make further matches at the next two lower concentrations of the sample and standard. Interpolation of matches may be made at half dilution steps. As the standard tubes contain known amounts of nisin, calculate the concentration of nisin in the sample solution. Obtain three readings of the solution and average them. Calculate the activity in terms of IU per gram of product.

Determination of sodium chloride: Transfer about 200 mg of the sample, accurately weighed, into a glass-stoppered flask containing 50 ml of water. Agitate the flask to dissolve the sample while adding 3 ml of nitric acid, 5 ml of nitrobenzene, 50.0 ml of standardized 0.1 N silver nitrate, and 2 ml of ferric ammonium sulfate TS. Shake the solution well, and titrate the excess silver nitrate with 0.1 N ammonium thiocyanate. The titration endpoint is indicated by the appearance of a red colour. Each ml of reacted 0.1 N silver nitrate is equivalent to 5.844 mg of NaCl. Calculate the percentage of sodium chloride in the sample taken by the equation:

Sodium chloride % (w/w) = 100 x 58.44(50 x A – V x B)/(W); where, A is the concentration of the silver nitrate solution; B is the concentration of the ammonium thiocyanate solution; V is the volume (ml) of the ammonium thiocyanate consumed; W is the weight of the sample (mg); and 58.44 is the formula weight of sodium chloride.

Nitrogen

Determine by Gas chromatography using the following conditions: Column: – material: stainless steel; – length: 2 m internal diameter: 2 mm; packing material: appropriate molecular sieve capable of absorbing molecules with diameters up to 0.5 nm; Carrier: – gas: helium (not less than 99.995 % (v/v) of He); flow: 40 ml/min; Detector: thermal conductivity detector; Injector: loop injector;

Column temperature: 50°; Detector temperature: 130°; Reference gas (a): ambient air; Reference gas (b): Nitrogen (not less than 99.999 % (v/v) of N_2, less than 1 ppm CO, less than 5 ppm O_2)

Procedure: Inject reference gas (a). Adjust the injected volumes and operating conditions so that the height of the peak due to nitrogen in the chromatogram is at least 35 % of full scale of the recorder. The assay is not valid unless the chromatograms obtained show a clear separation of oxygen and nitrogen. Inject the gas to be examined and the reference gas (b). In the chromatogram obtained with the gas to be examined, the area of the principal peak is at least 99.0 % of the area of the principal peak in the chromatogram obtained with reference gas (b).

Nitrous oxide

Using a gas sampling valve inject standard gas taken from the liquid phase and record the area under the peak of nitrous oxide. Adjust the injection volume and operating conditions so that a good quantifiable peak for nitrous oxide is obtained (not less than the 35% of the full scale when using an integrator). Record area under the peak of nitrous oxide for the standard. Inject the sample gas taken from the liquid phase and record the area. Calculate the purity of sample gas from the peak areas of standard and sample, and purity of certified nitrous oxide standard.

Nordihydrogu aiaretic acid

Weigh 1.00 g of the sample. Dilute with methanol so that the final concentration will be 1 mg of the sample per 100 ml of solution. Read the absorbance at 284 nm in a 1 cm quartz cell.

Octyl gallate

Weigh accurately about 0.2 g of the dried sample into a 400-ml beaker. Add 150 ml of water and heat to boiling. Then with constant and vigorous stirring add 50 ml of bismuth nitrate TS (II). Continue stirring for a few min more until precipitation is complete, then allow the solution to cool to room temperature. Filter the yellow precipitate on a weighed sintered glass crucible, wash first with cold 0.05N nitric acid and then with ice-cold water, until free from acid. Dry at 110^0 to constant weight.

Oxygen
Place a sufficient quantity of ammonium chloride-ammonium hydroxide solution, prepared by mixing equal volumes of water and ammonium hydroxide and saturating with ammonium chloride at room temperature, in test apparatus composed of a calibrated 100-ml burette, provided with two-way stopcock, a gas absorption pipette, and a levelling bulb, both of suitable capacity and all suitably interconnected. Fill the gas absorption pipette with metallic copper in the form of wire coils, wire mesh, or other suitable configuration. Eliminate all gas bubbles from the liquid in the test apparatus. Activate the test solution by performing two to three tests that are not for record purposes. Fill the calibrated burette, all interconnecting tubing, both stopcock opening, and the intake tube with liquid. Draw 100.0 ml of oxygen into the burette by lowering the levelling bulb. Open the stopcock to the absorption pipette, and force the oxygen into the absorption pipette by raising the levelling bulb. Agitate the pipette to provide frequent and intimate contact of the liquid, gas, and copper. Continue agitation until no further diminution in volume occurs. Draw the residual gas back into calibrated burette, and measure its volume: not more than 1.0 ml of gas remains.

Octanoic acid
Determine using an appropriate gas chromatographic technique. The selection of sample size and method of sample preparation may be based on AOCS Method Cc 1-62 and Ce 1f-96 and follow the principles of the method described in FNP 5 "Instrumental methods". The percentage of octanoic acid is the area percent of the methyl octanoate peak. Decanoic acid may be determined using the same gas chromatographic technique.

O-phenyl
Weigh 2.000 g of o-phenylphenol, dissolve in 10 ml of 10% sodium hydroxide solution by

phenol

warming and dilute to 500.0 ml. Pipette 25.0 ml into a 250-ml iodine flask and add 30.0 ml of 0.1 N bromide-bromate TS and 50 ml of anhydrous methanol. Place the stopper in the flask and add 10 ml of dilute (1 to 1) hydrochloric acid to the well. Raise the stopper slightly to allow the acid to flow down the sides inside the flask, but retain a small amount of the acid in the well to act as a seal. Mix the contents by swirling and allow it to react for exactly 30 sec at 25±5o. Immediately add 10 ml of 20% potassium iodide solution to the well and allow it to drain into the assay flask as for the acid. Mix well, allow the solution to stand for 5 min, shaking occasionally. Wash the stopper and the sides of the flask with water. Titrate the liberated iodine with 0.1 N sodium thiosulfate adding starch TS as the endpoint is approached. Each ml of 0.1 N bromide-bromate TS consumed is equivalent to 4.255 mg of $C_{12}H_{10}O$.

Phosphoric acid

Weigh 1.00 g of the sample into a glass-stoppered flask, dilute with about 100 ml of water, add 0.5 ml of thymolphthalein TS, and titrate with 1N sodium hydroxide. Each ml of 1N sodium hydroxide is equivalent to 0.049 g of H_3PO_4.

Polydextroses

Phenol Solution: Add 20 ml of water to 80 g of phenol.

Glucose Standard Solutions: Weigh accurately 100 mg of alpha-D-glucose (minimum 97% purity) into a 500-ml volumetric flask and make up to volume with distilled water. Dilute five aliquots of the solution with distilled water to obtain the following concentrations of standard: 50, 40, 20, 10 and 5 µg/ml.

Standard Curve: Run each analysis in triplicate. On a daily basis, pipet 2.0 ml of each of the Glucose Standard Solutions into 4-dram (14.8 ml) acetone-free screw-cap vials. Add 0.12 ml of the phenol solution and mix gently. Uncap vials and add rapidly 5 ml of sulfuric acid TS.

Immediately recap the vials and shake vigorously.

<u>CAUTION</u>: Rubber gloves and safety shield should be used in the sulfuric acid addition step. Let the vials stand at room temperature for 45 min. Determine absorbances at 490 nm in a suitable spectrophotometer, using a Phenol Solution-sulfuric acid mixture as a blank in the reference cell. Plot mean absorbances versus concentrations in μg/ml.

<u>Procedure:</u> In triplicate, weigh accurately about 250 mg of the sample into a 250-ml volumetric flask and make up to volume with distilled water. Transfer a 10 ml aliquot to a 250-ml volumetric flask and make up to volume with distilled water.

Ponceau 4r	Proceed by Titration with Titanous Chloride, using the following: Weight of sample: 0.7-0.8 g; Buffer: 10 g sodium citrate; Weight (D) of colouring matters equivalent to 1.00 ml of 0.1 N $TiCl_3$: 15.78 mg
Potassium acetate	Dissolve about 200 mg of the dried sample, accurately weighed, in 25 ml of glacial acetic acid. Add 2 drops of crystal violet TS, and titrate with 0.1 N perchloric acid in glacial acetic acid. Perform a blank determination, and make any necessary correction. Each ml of 0.1 N perchloric acid is equivalent to 9.814 mg of $C2H3KO2$
Potassium alginate	Proceed as directed under Carbon Dioxide Determination by Decarboxylation. Each ml of 0.25 N sodium hydroxide consumed is equivalent to 5.5 mg of carbon dioxide (CO_2) or 29.75 mg of potassium alginate (equivalent weight 238).
Potassium benzoate	Weigh to the nearest 0.1 mg, 2.5 to 3 g of the dried sample, and transfer to a 250-ml Erlenmeyer flask. Add 50 ml of water to dissolve the sample. Neutralize the solution, if necessary, with 0.1N hydrochloric acid, using phenolphthalein TS as indicator. Add 50 ml of ether and a few drops of

bromophenol blue TS and titrate with 0.5N hydrochloric acid, shaking constantly the flask, until the colour of the indicator begins to change. Transfer the lower aqueous layer to another flask. Wash the ethereal layer with 10 ml of water, and add the washing and an additional 20 ml of ether to the separated aqueous layer. Complete the titration with the 0.5N hydrochloric acid, shaking constantly the flask. Each ml of 0.5N hydrochloric acid is equivalent to 80.11 mg of $C_7H_5KO_2$.

Potassium carbonate	Weigh accurately about 1 g of the dried sample. Dissolve carefully in 50 ml of 1 N sulfuric acid, add methyl orange TS and titrate the excess acid with 1N sodium hydroxide. Each ml of 1 N sulfuric acid is equivalent to 69.11 mg of K2CO3.
Potassium chloride	Dissolve about 250 mg of the dried sample, accurately weighed in 50 ml of water in a glass-stoppered flask. Add, while agitating, 50 ml of 0.1 N silver nitrate, 3 ml of nitric acid, and 5 ml of nitrobenzene, shake vigorously, add 2 ml of ferric ammonium sulfate TS, and titrate the excess silver nitrate with 0.1 N ammonium thiocyanate. Each ml of 0.1 N silver nitrate is equivalent to 7.456 mg of KCl.
Potassium dihydrogen phosphate	Transfer about 5 g of the dried sample, accurately weighed, into a 250-ml beaker. Add 100 ml of water and 5 ml of 1 N hydrochloric acid, and stir until the sample is completely dissolved. Place the electrodes of a suitable pH meter in the solution, and slowly titrate the excess acid, stirring constantly, with 1 N sodium hydroxide to the inflection point occurring at about pH 4. Record the buret reading, and calculate the volume (A), if any, of 1 N hydrochloric acid consumed by the sample. Continue the titration with 1 N sodium hydroxide until the inflection point occurring at about pH 8.8 is reached, record the buret reading, and calculate the volume (B) of 1 N sodium hydroxide required in the titration between the two inflection points (pH 4 and pH 8.8). Each ml of

the volume (B) - (A) of 1 N sodium hydroxide is equivalent to 136.1 mg of KH_2PO_4

Potassium gluconate	Transfer about 175 mg of the sample, accurately weighed, into a clean, dry 200-ml Erlenmeyer flask, add 75 ml of glacial acetic acid and dissolve by heating on a hot plate. Cool, add quinaldine red TS, and titrate with 0.1 N perchloric acid in glacial acetic acid, using a 10-ml microburet, to a colourless end point. Each ml of 0.1 N perchloric acid is equivalent to 23.42 mg of $C_6H_{11}KO_7$.
Potassium hydrogen carbonate	Dissolve about 4 g of the sample, accurately weighed, in 25 ml of water, add methylorange TS and titrate with 1 N sulfuric acid. Each ml of 1 N sulfuric acid is equivalent to 100.1 mg of $KHCO_3$
Potassium hydroxide	Dissolve about 1.5 g of the sample, accurately weighed, in 40 ml of recently boiled and cooled water, cool to 15o, add phenolphthalein TS and titrate with 1 N sulfuric acid. At the discharge of the pink colour, record the volume of acid required, then add methyl orange TS and continue to titrate to a persistent pink colour. Record the total volume of acid required for the titration. Each ml of 1 N sulfuric acid is equivalent to 56.11 mg of total alkali, calculated as KOH
Quinine hydrochloride	Dissolve about 150 mg of the sample, accurately weighed, in 20 ml of acetic anhydride, add 2 drops of malachite green TS and 5.5 ml of mercuric acetate TS and titrate with 0.1 N perchloric acid from a micro-buret to a yellow endpoint. Perform a blank determination and correct the sample titre as necessary. Each ml of 0.1 N perchloric acid is equivalent to 18.04 mg of $C_{20}H_{24}N_2O_2 \cdot HCl$.
Quinine sulfate	Dissolve about 200 mg of the sample, accurately weighed, in 20 ml of acetic anhydride, add 2 drops of 1% malachite green TS and titrate with 0.1N perchloric acid from a microburet to a yellow endpoint. Perform a blank determination. Each ml of 0.1 N perchloric acid is equivalent to

24.90 mg of $(C_{20}H_{24}N_2O_2)_2 \cdot H_2SO_4$

Quinoline yellow	Proceed as directed under Total Content by Spectrophotometry Solvent: pH 7 phosphate buffer. Dilution of solution A: 10 ml to 250 ml. Absorptivity (a): 86.5. Approximate wavelength of maximum absorption: 415 nm. Determination of the percentages of di-, mono- and trisulfonates in Quinoline Yellow prepared from 2-(2-quinolyl)-1,3-indandione (only): Use the HPLC conditions prescribed in the Determination of Organic Compounds other than Colouring Matters with a sample solution of concentration 0.05% in HPLC Solvent A in place of the sample solution of concentration 1%. Express the results as percentages of the Total colouring matters present.
Red 2g	Proceed Titration with Titanous Chloride using the following: Weight of sample: 0.6-0.7 g; Buffer: 15 g sodium hydrogen tartrate; Weight: (D) of colouring matters equivalent to 1.00 ml of 0.1 N $TiCl_3$: 12,74 mg
Riboflavin	Carry out the assay in subdued light. In a brown glass 500 ml volumetric flask, suspend 65.0 mg of the sample in 5 ml of water, ensuring that it is completely wetted, and dissolve in 5 ml of 2 N sodium hydroxide solution. As soon as dissolution is complete, add 100 ml of water and 2.5 ml of glacial acetic acid and dilute to 500.0 ml with water. Place 20.0 ml of this solution in a brown glass 200 ml volumetric flask, add 3.5 ml of a 1.4% w/v solution of sodium acetate and dilute to 200.0 ml with water. Measure the absorbance (A) at the maximum at 444 nm.
Riboflavin 5'-phosphate sodium	Carry out the assay in subdued light. In a brown glass 500 ml volumetric flask, dissolve 100 mg of the sample in 100 ml of water and add 2.5 ml of glacial acetic acid and dilute to 500.0 ml with water. Place 20.0 ml of this solution in a brown glass 200 ml volumetric flask, add 3.5 ml of a 1.4% w/v solution of sodium acetate and dilute to 200.0 ml with water. Measure the absorbance

(A) at the maximum at 444 nm.

Riboflavin from bacillus subtilis

Carry out the assay in subdued light. In a brown-glass 500-ml volumetric flask, suspend 65.0 mg of the sample in 5 ml of water, ensuring that it is completely wetted, and dissolve in 5 ml of 2 N sodium hydroxide solution. As soon as dissolution is complete, add 100 ml of water and 2.5 ml of glacial acetic acid and dilute to 500.0 ml with water. Place 20.0 ml of this solution in a brown glass 200-ml volumetric flask, add 3.5 ml of a 1.4% w/v solution of sodium acetate and dilute to 200.0 ml with water. Measure the absorbance (A) at the maximum, 444 nm.

Sodium nitrite

Weigh, to the nearest mg, 1 g of the dried sample. Transfer to a 100 ml volumetric flask and dissolve in water diluting to the mark. Pipette 10.0 ml of this solution into a mixture of 50.0 ml of 0.1N potassium permanganate, 100 ml of water and 5 ml of sulfuric acid, keeping the tip of the pipette well below the surface of the liquid. Warm the solution to 40o, allow it to stand for 5 min, and add 25.0ml of 0.1N oxalic acid. Heat the mixture to about 80o and titrate with 0.1N potassium permanganate.

Sodium o-phenylphenol

Weigh 3.100 g of sodium o-phenylphenol, dissolve in water, adding a few drops of 10% sodium hydroxide solution if necessary to clear any turbidity, and dilute to 500.0 ml with water. Pipette 25.0 ml into a 250- ml iodine flask, and add 30.0 ml of 0.1 N bromide-bromate TS and 50 ml of anhydrous methanol. Place stopper in the flask and add 10 ml of dilute (1 to 1) hydrochloric acid to the well. Raise the stopper slightly to allow the acid to flow down the sides of the flask, but retain a small amount of the acid in the well to act as a seal. Mix the contents by swirling and allow it to react for exactly 30 sec at 25±5o. Immediately add 10 ml of 20% potassium iodide solution to the well and allow it to drain into the flask. Mix well and allow the solution to stand for 5 min, shaking occasionally. Wash the stopper and the sides of the flask with water and titrate the liberated iodine with 0.1 N sodium thiosulfate adding starch TS as the endpoint is approached. Each ml of 0.1 N bromide-bromate TS consumed is equivalent to 6.608 mg of $C_{12}H_9ONa \cdot 4H_2O$.

Sodium percarbonate	Using a measuring cylinder, carefully add 100 ml of sulphuric acid solution (3.6 N) to a 600-ml beaker. Weigh accurately about 4 g of sample on to a tared watchglass. Let the weight of sample be W g. Place the watchglass in the beaker, cover the beaker with a clockglass and swirl to dissolve the sample. Transfer the solution to a 500 ml volumetric flask. Rinse the clockglass and the wall of the beaker with demineralised water and add all the washings to the volumetric flask. Dilute to volume with demineralised water and mix well. Immediately titrate a portion of this solution. Add 100 ml of sulphuric acid solution (3.6 N) to a 600 ml conical flask and add potassium permanganate solution (0.1 N) dropwise to the appearance of a faint permanent pink colour. Using a safety pipette, add 25.0 ml of sample solution and mix well. Titrate with potassium permanganate solution (0.1 N) to the reappearance of the faint permanent pink colour. Let the titration obtained be A ml.
Sodium polyphosphates- glassy	Transfer about 800 mg of the sample, accurately weighed, into a 400-ml beaker. Add 100 ml of water and 25 ml of nitric acid, cover with a watch glass, and boil for 10 min on a hot plate. Rinse any condensate from the watch glass into the beaker; cool the solution to room temperature; transfer it quantitatively to a 500-ml volumetric flask; dilute to volume with water; and mix thoroughly. Pipet 20.0 ml of this solution into a 500-ml Erlenmeyer flask, add 100 ml of water, and heat just to boiling. Add with stirring 50 ml of quimociac TS, then cover with a watch glass, and boil for 1 min in a well-ventilated hood. Cool to room temperature, swirling occasionally

	while cooling, then filter through a tared, sintered-glass filter crucible of medium porosity, and wash with five 25-ml portions of water. Dry at about 225o for 30 min, cool, and weigh. Each mg of precipitate thus obtained is equivalent to 32.074 µg of P2O5.
Sodium propionate	Weigh, to the nearest mg, 3 g of the sample previously dried at 105° for 1 h, into a distillation flask and add 200 ml of 50% phosphoric acid. Boil for 2 h and collect the distillate. During distillation keep the volume in the flask at about 200 ml by adding water using a dropping funnel. Titrate the distillate with 1N sodium hydroxide using phenolphthalein TS as indicator. Each ml of 1N sodium hydroxide corresponds to 96.06 mg of $C_3H_5NaO_2$.
Sodium saccharin	Dissolve about 0.3 g of previously dried sample, accurately weighed, in 20 ml of glacial acetic acid. Add 2 drops of crystal violet-glacial acetic acid TS as indicator, and titrate with 0.1 N perchloric acid. End-point is where violet colour of the solution changes to green, via blue. Perform a blank determination, and make any necessary correction. Each ml of 0.1 N perchloric acid is equivalent to 20.52 mg of C7H4NNaO3S.
Sodium sesquicarbonate	$NaHCO_3$: Weigh accurately about 8.5 g of the sample in a 250-ml flask, and dissolve it in 50 ml of carbon dioxide-free water. Titrate with 1 N sodium hydroxide until a drop of the solution, when added to a drop of a 1-in-10 solution of silver nitrate TS on a spot plate, produces a dark brown colour. Each ml of 1 N sodium hydroxide is equivalent to 84.01 mg of $NaHCO_3$. Na_2CO_3: Weigh accurately about 4.2 g of the sample in a 250 ml flask, and dissolve it by adding 100 ml of water. Add 3 drops of methyl orange TS and titrate with 1 N sulfuric acid, stirring vigorously near the end point to expel carbon dioxide. Each ml of 1 N sulfuric acid is equivalent to 30.99 mg of Na_2O. C

Sodium sulfate
: Weigh accurately about 0.5 g of the dried sample, dissolve in 200 ml of water, add 1 ml of hydrochloric acid and heat to boiling. Gradually add, in small portions and while stirring constantly, an excess of hotbarium chloride TS (about 10 ml), and heat the mixture on a steam bath for 1 h. Collect the precipitate on a filter, wash until free from chloride, dry, ignite and weigh. The weight of the barium sulfate so obtained, multiplied by 0.6086 corresponds to the equivalent amount of Na2SO4.

Sodium sulfite
: Weigh 250 mg of the sample, add to 50.0 ml of 0.1 N iodine in a glass-stoppered flask, and stopper the flask. Allow to stand for 5 min, add 1 ml of hydrochloric acid and titrate the excess iodine with 0.1 N sodium thiosulfate, adding starch TS as the indicator. Each ml of 0.1 N iodine is equivalent to 6.302 mg of Na_2SO_3.

Sodium thiocyanate
: Dissolve about 1 g of the sample, accurately weighed in water in a volumetric flask and dilute to 100 ml. Place this solution in a burette. Place in a 250-ml conical flask 10 ml of 0.1N silver nitrate, add 5 ml of dilute nitric acid TS and 2 ml of ferric ammonium sulfate TS. Titrate with the sample solution until a reddish yellow colour is obtained. 10 ml of $AgNO_3$ is equivalent to 84.1 mg of NaSCN.

Sodium thiosulfate
: Dissolve about 0.5 g of the dried sample, accurately weighed, in 30 ml of water and titrate with 0.1 N iodine solution using starch TS as the indicator. Each ml of 0.1 N iodine is equivalent to 15.81 mg of $Na_2S_2O_3$.

Sorbic acid
: Dissolve about 0.25 g of the sample, accurately weighed, in 50 ml of anhydrous methanol previously neutralized with 0.1 N sodium hydroxide, add phenolphthalein TS, and titrate with 0.1 N sodium hydroxide to the first pink colour which persists for at least 30 sec. Each ml of 0.1 N

sodium hydroxide is equivalent to 11.21 mg of $C_6H_8O_2$.

Sorbitan monolaurate	Transfer about 25 g of the sample, accurately weighed, into a 500-ml round-bottom flask, add 250 ml of alcohol and 7.5 g of potassium hydroxide, and mix. Connect a suitable condenser to the flask, reflux the mixture for 1 to 2 h, and then transfer to an 800-ml beaker, rinsing the flask with about 100 ml of water and adding it to the beaker. Heat on a steam bath to evaporate the alcohol, adding water occasionally to replace the alcohol, and evaporate until the odour of alcohol can no longer be detected. Adjust the final volume to about 250 ml with hot water. Neutralize the soap solution with dilute sulfuric acid (1 in 2), add 10% in excess, and heat, while stirring, until the fatty acid layer separates. Transfer the fatty acids to a 500-ml separator, wash with three or four 20-ml portions of hot water to remove polyols, and combine the washings with the original aqueous polyol layer from the saponification. Extract the combined aqueous layer with three 20-ml portions of petroleum ether, add the extracts to the fatty acid layer, evaporate to dryness in a tared dish, cool and weigh. Neutralize the polyol solution with a 1 in 10 solution of potassium hydroxide to pH 7 using a suitable pH meter. Evaporate this solution to a moist residue, and separate the polyols from the salts by several extractions with hot alcohol. Evaporate the alcohol extracts on a steam bath to dryness in a tared dish, cool, and weigh. Avoid excessive drying and heating.
Sorbitan monooleate	Transfer about 25 g of the sample, accurately weighed, into a 500-ml round-bottom flask, add 250 ml of alcohol and 7.5 g of potassium hydroxide, and mix. Connect a suitable condenser to the flask, reflux the mixture for 1 to 2 h, and then transfer to an 800-ml beaker, rinsing the flask with about 100 ml of water and adding it to the beaker. Heat on a steam bath to evaporate the alcohol, adding water occasionally to replace the alcohol, and evaporate until the odour of alcohol can no

longer be detected, Adjust the final volume to about 250 ml with hot water. Neutralize the soap solution with dilute sulfuric acid (1 in 2), add 10% in excess, and heat, while stirring, until the fatty acid layer separates. Transfer the fatty acids to a 500-ml separator, wash with three or four 20-ml portions of hot water to remove polyols, and combine the washings with the original aqueous polyol layer from the saponification. Extract the combined aqueous layer with three 20-ml portions of petroleum ether, add the extracts to the fatty acid layer, evaporate to dryness in a tared dish, cool, and weigh. Neutralize the polyol solution with a 1 in 10 solution of potassium hydroxide to pH 7 using a suitable pH meter. Evaporate this solution to a moist residue, and separate the polyols from the salts by several extractions with hot alcohol. Evaporate the alcohol extracts on a steam bath to dryness in a tared dish, cool, and weigh. Avoid excessive drying and heating. Assay another 25 g sample by the Sorbitan Ester Content procedure to determine percent sorbitan ester.

Sorbitan tristearate

Transfer about 25 g of the sample, accurately weighed, into a 500-ml round-bottom flask, add 250 ml of alcohol and 7.5 g of potassium hydroxide, and mix. Connect a suitable condenser to the flask, reflux the mixture for 1 to 2 h, and then transfer to an 800-ml beaker, rinsing the flask with about 100 ml of water and adding it to the beaker. Heat on a steam bath to evaporate the alcohol, adding water occasionally to replace the alcohol, and evaporate until the odour of alcohol can no longer be detected. Adjust the final volume to about 250 ml with hot water. Neutralize the soap solution with dilute sulfuric acid (1 in 2), add 10% in excess, and heat, while stirring, until the fatty acid layer separates. Transfer the fatty acids to a 500-ml separator, wash with three or four 20-ml portions of hot water to remove polyols, and combine the washings with the original aqueous polyol layer from the saponification. Extract the combined aqueous layer with three 20-

ml portions of petroleum ether, add the extracts to the fatty acid layer, evaporate to dryness in a tared dish, cool and weigh. Neutralize the polyol solution with a 1 in 10 solution of potassium hydroxide to pH 7 using a suitable pH meter. Evaporate this solution to a moist residue, and separate the polyols from the salts by several extractions with hot alcohol. Evaporate the alcohol extracts on a steam bath to dryness in a tared dish, cool, and weigh. Avoid excessive drying and heating. Assay another 25 g sample by the Sorbitan Ester Content procedure to determine percent sorbitan ester.

Stannous chloride
Transfer about 2 g of sample accurately weighed, to a 250 ml volumetric flask, dissolve in 25 ml of hydrochloric acid, dilute to volume with water, and mix well. Transfer 50 ml of this solution to a 500 ml conical flask, and add 5 g of potassium sodium tartrate, and then a cold saturated solution of sodium bicarbonate until de solution is alkaline to litmus paper. Titrate at once with 0.1 N iodine using starch TS as the indicator. Each ml of 0.1N iodine consumed is equivalent to 11.28 mg of $SnCl_2 \cdot 2H_2O$.

(Note: Stannous salts are readily susceptible to oxidation, yet the method of assay does not take account of this. Distilled water contains dissolved oxygen, therefore water used in the method of assay should be "oxygen free"; this may be achieved by purging the water with nitrogen or carbon dioxide or by boiling the air out. In addition to this the iodine solution used in the determination should be free from dissolved oxygen; ideally the iodine solution should be stored in a self-filling apparatus under carbon dioxide.)

Sorbitol syrup

Determine the sorbitol content of the sample using *liquid chromatography. Apparatus:* Liquid chromatograph (HPLC); Detection: differential refractometer maintained at constant temperature Integrator recorder. Column: AMINEX HPX 87 C (or equivalent resin in calcium form), length 30 cm, internal diameter 9 mm; Eluent: double distilled degassed water (filtered through Millipore membrane filter 0.45 μm). Chromatographic conditions; Column temperature: 85 ± 0.5^0; Eluent flow rate: 0.5 ml/min

Standard preparation: Dissolve an accurately weighed quantity of sorbitol in water to obtain a solution having known concentration of about 10.0 mg of sorbitol per ml.

Sample preparation: Transfer about 1 g of the sample accurately weighed to a 50 ml volumetric flask, dilute with water to volume and mix.

Procedure: Separately inject equal volumes (about 20 μl) of the sample preparation and the standard preparation into the chromatograph. Record the chromatograms and measure the responses of each polyol peak. Calculate separately the quantity, in mg, of sorbitol in the portion of sample taken by the following formula: 50 * C * RU*RS; where, C = concentration, in mg per ml, of sorbitol in the standard preparation; RU = the peak response of the sample preparation; RS = the peak response of the standard preparation.

Stearyl tartrate

Add exactly 50 ml of 0.1N ethanolic potassium hydroxide solution to the neutralized solution obtained above from the determination of the acid value and bring to a steady boil for 2 min. Cool the solution and titrate the excess of potassium hydroxide with B ml of 0.1N hydrochloric acid. Perform a blank determination (= A ml of 0.1N hydrochloric acid).

Sucralose

Chromatographic system: Fit a high pressure liquid chromatograph, operated at room temperature, with a radial compression module containing a 10 cm 5 μm C18 reverse phase column. The mobile phase is maintained at a pressure and flow rate (typically 1.5 ml/min) capable of giving the required elution time (see System Suitability Test). An ultraviolet detector that monitors absorption at 190 nm, or a refractive index detector, is used.

Mobile Phase: Add 150 ml of acetonitrile (HPLC grade, far UV, filtered through a 0.45 μm Millipore filter or equivalent) to 850 ml of water (glass distilled, filtered through a 0.45 μm Millipore filter or equivalent). Mix and de-gas thoroughly.

Standard Solution: Weigh accurately about 250 mg of sucralose Reference Standard into a 25 ml volumetric flask. Dissolve and make up to volume using the Mobile Phase. Filter the solution through a 0.45 μm Millipore filter or equivalent. Record the weight of Reference Standard as Ws.

Test solution: Weigh accurately about 250 mg of sample into a 25 ml volumetric flask. Dissolve and make up to volume using the mobile phase. Filter the solution through a 0.45 μm Millipore filter or equivalent. Record the weight of sample as Wt.

System Suitability Test: Inject duplicate 20 μl portions of Standard Solution into the chromatograph. The retention time of the sucralose should be approximately 9 min. (NOTE: The retention time quoted is appropriate for a 10 cm 5 μm Rad-Pak C18 column. If a column of a different make or length is used it may be necessary to adjust the proportion of acetonitrile in the eluent to obtain the required retention time). The co-efficient of variation (100 x standard deviation divided by mean peak area) for the peak areas should not exceed 2%.

Procedure: Analyse the Test Solution under the conditions described above, making duplicate 20 μl injections, and calculate the mean peak area. Calculate the percentage purity from the relative peak areas of the Test (At) and Standard (As) Solutions according to the following formula: % Purity = At* Ws / (As* Wt); Calculate the percentage purity on a water-free and methanol-free basis using the values obtained in the tests for water and methanol.

Sucroglyceride s	Determine by *high pressure liquid chromatography* (see Volume 4) using the following conditions:
	Sample preparation: Add about 250 mg of the sample, accurately weighed to a 50 ml volumetric flask. Dilute to volume with tetrahydrofuran, and mix. Filter through a 0.5-μm membrane filter.
	Procedure: Inject 100 μl of the sample into the pre-stabilized high pressure liquid chromatograph.
	Conditions: Column: Styrene-divinylbenzene copolymer for gel permeation chromatography (TSK-GEL G2000 (Supelco) or equivalent); Mobile phase: HPLC-grade degassed tetrahydrofuran; Flow rate: 0.7 ml/min; Detector: Refractive index detector; Temperatures: Column: 38°; Detector: 38°; Record the chromatogram for about 90 min. Calculate the percentage of sucrose ester content in the sample.
Sulfuric acid	Transfer a 1-ml sample into a small, tared, glass-stoppered conical flask, insert the stopper, weigh accurately, and cautiously add about 30 ml of water. Cool the mixture, add methyl orange TS, and titrate with 1 N sodium hydroxide. Each ml of 1 N sodium hydroxide is equivalent to 49.04 mg of H_2SO_4.
Sulfur dioxide	Subtract from 100 the percentages of non-volatile residue and of water, as determined herein, to obtain the percentage of SO_2.

D-tagatose

Determined by liquid chromatography: Preparation of sample solution: Weigh accurately about 50 mg of dry sample into a 10-ml volumetric flask and add about 8 ml of purified, deionized water. Bring sample to complete dissolution and dilute to mark with purified deionized water. Filter through a 0.2 μm filter. Preparation of reference solution: Use dry standard D-tagatose. Prepare a solution of the reference material as described for the sample solution.

Apparatus: Liquid chromatograph equipped with a refractive index detector and an integrator. Conditions: Column: Biorad Aminex HPX-87C (length 30 cm, diameter; 7.8 mm, particle size 9 μm) or equivalent; Column temperature: 85°; Mobile phase: Deionized water with 50 ppm calcium acetate; Flow rate: 0.6 ml/min ; Injection volume: 20 μl;

Procedure: Separately inject equal volumes of the sample solution and the reference solution into the chromatograph. Record the chromatograms and measure the response of D-tagatose peak.

Tagetes extract

Accurately weigh 0.5 - 1.5 g of the sample into a 50 ml volumetric flask. Add about 30 ml of a 1:1 mixture of cyclohexane and ethanol (96%) and swirl gently until the sample is dissolved. Fill to the mark with cyclohexane/ethanol mixture and mix well. Pipette 0.200 ml of the solution into a 25 ml volumetric flask. Make up to mark with cyclohexane/ethanol and mix well. Measure the absorbance at maximum of about 444 nm using cyclohexane/ethanol mixture as blank. The absorbance should be between 0.2 and 0.8, otherwise an appropriate dilution must be prepared.

Tannic acid

Sample test: Accurately weigh about 2.0 g of the sample (W), transfer to a 500-ml volumetric flask, add water to dissolve and make up to the volume with water. Transfer 100 ml of this solution into a 300 ml Erlenmeyer flask and add 7.2 g of Hide Powder (a suitable grade is available from L.H. Lincoln & Son, Inc., Tanning Materials, Coudersport, Pennsylvania, 16915

USA). Shake the flask for 20 min. Let stand for 10 min. and filter through a G4-filter. The filtrate shall be clear. Pipette 50 ml of the filtrate into a tared crystallizing dish. Evaporate to dryness on a steam bath and heat in an oven at 105° for 1 h. Cool in a desiccator and weigh (a).

Blank test: On each lot of Hide Powder (a suitable grade of Hide Powder may be obtained from L.H. Lincoln & Son, Inc., Tanning Materials, Coudersport, Pennsylvania, 16915 USA) a blank test has to be carried out. Weigh 7.2 g of Hide Powder EFT into a 300 ml Erlenmeyer containing 100 ml water. Proceed as directed for the Sample test, beginning with "Shake the flask for 20 min...". Cool in a desiccator and weigh (b).

L(+)-tartaric acid	Weigh accurately about 2 g of the dried sample, dissolve in 40 ml of water, add phenolphthalein TS, and titrate with 1 N sodium hydroxide. Each ml of 1 N sodium hydroxide is equivalent to 75.04 mg of C4H6O6.
Dl-tartaric acid	Weigh accurately about 2 g of the dried sample, dissolve it in 40 ml of water, add phenolphthalein TS, and titrate with 1 N sodium hydroxide. Each ml of 1 N sodium hydroxide is equivalent to 75.04 mg of C4H6O6.
Tartrazine	Proceed as directed under Total Content by Titration with Titanous Chloride (see Volume 4), using the following: Weight of sample: 0.6-0.7 g; Buffer: 15 g sodium hydrogen tartrate; Weight (D) of colouring matters equivalent to 1.00 ml of 0.1 N TiCl3: 13.56 mg
Tertiary butylhydroqui none	Transfer about 170 mg of the sample, previously ground to a fine powder ASSAY and accurately weighed, into a 250-ml wide-mouth conical flask, and dissolve in 10 ml of methanol. Add 150 ml of water, 1 ml of N sulfuric acid, and 4 drops of diphenylamine indicator (3 mg of pdiphenylaminesulfonic acid, sodium salt, per ml of 0.1 N sulfuric acid), and titrate with 0.1 N

ceric sulfate to the first complete colour change from yellow to red-violet. Record the volume, in ml, of 0.1 N ceric sulfate required as V.

Tetrapotassiu m pyrophosphate	Dissolve about 600 mg of the sample, accurately weighed, in 100 ml of water in 400-ml beaker, and adjust the pH of the solution to exactly 3.8 with hydrochloric acid, using a pH meter. Add 50 ml of a 1 in 8 solution of zinc sulfate (125 g of $ZnSO_4 \cdot 7H_2O$ dissolved in water, diluted to 1000 ml, filtered, and adjusted to pH 3.8) and allow to stand for 2 min. Titrate the liberated acid with 0.1 N sodium hydroxide until a pH of 3.8 is again reached. After each addition of sodium hydroxide near the end-point, time should be allowed for any precipitated zinc hydroxide to redissolve. Each ml of 0.1 N sodium hydroxide is equivalent to 16.52 mg of $K_4P_2O_7$.
Tetrasodium pyrophosphate	Dissolve an accurately weighed quantity of the sample, equivalent to about 500 mg of anhydrous $Na_4P_2O_7$, in 100 ml of water in a 400-ml beaker. Adjust the pH of the solution to 3.8 with hydrochloric acid, using a pH meter, then add 50 ml of a 1 in 8 solution of zinc sulfate (125 g of $ZnSO_4 \cdot 7H_2O$ dissolved in water, diluted to 1000 ml, filtered, and adjusted to pH 3.8) and allow to stand for 2 min. Titrate the liberated acid with 0.1 N sodium hydroxide until a pH of 3.8 is again reached. After each addition of sodium hydroxide near the end-point, time should be allowed for any precipitated zinc hydroxide to redissolve. Each ml of 0.1 N sodium hydroxide is equivalent to 13.30 mg of $Na_4P_2O_7$
Thiodipropioni c acid	Dissolve 0.350 g of the sample in 40 ml of water, add phenolphthalein TS and titrate with 0.1 N sodium hydroxide to the first appearance of a faint pink colour that persists for at least 30 sec. Each ml of 0.1 N sodium hydroxide is equivalent to 8.910 mg of $C_6H_{10}O_4S$.

Titanium dioxide

Accurately weigh about 150 mg of the sample, previously dried at 105° for 3 hours, and transfer into a 500-ml conical flask. Add 5 ml of water and shake until a homogeneous, milky suspension is obtained. Add 30 ml of sulfuric acid and 12 g of ammonium sulfate, and mix. Initially heat gently, then heat strongly until a clear solution is obtained. Cool, then cautiously dilute with 120 ml of water and 40 ml of hydrochloric acid, and stir. Add 3 g of aluminium metal, and immediately insert a rubber stopper fitted with a U-shaped glass tube while immersing the other end of the U-tube into a saturated solution of sodium bicarbonate contained in a 500-ml wide-mouth bottle, and generate hydrogen. Allow to stand for a few minutes after the aluminium metal has dissolved completely to produce a transparent purple solution. Cool to below 50°in running water, and remove the rubber stopper carrying the U-tube. Add 3 ml of a saturated potassium thiocyanate solution as an indicator, and immediately titrate with 0.1 N ferric ammonium sulfate until a faint brown colour that persists for 30 seconds is obtained. Perform a blank determination and make any necessary correction. Each ml of 0.1 N ferric ammonium sulfate is equivalent to 7.990 mg of TiO_2.

Trehalose

Determine by liquid chromatography using the following conditions. Sample preparationWeigh accurately about 3g of dry sample into a 100ml volumetric flask and add about 80 ml of deionized water. Bring sample to complete dissolution and dilute to mark with deionized water. Filter through a 0.45 micron filter. Preparation of standard solutionDissolve accurately weighed quantities of dry standard reference trehalose (available from Hayashibara Co., Ltd, 2-3 Shimoishii1-chome, Okayama 700, Japan) in water to obtain a solution having known concentration of about 30 mg of trehalose per ml. ApparatusLiquid chromatograph equipped with a refractive index detector and an integrating recorder.

Conditions Column and packing : Shodex Ionpack KS-801 or equivalent -length : 300mm; -diameter : 10mm; -temperature : 50°C; Mobile phase : water; Flow rate : 0.4 ml/min; Injection volume : 8µl

Procedure: Separately inject equal volumes of the sample solution and the standard solution into the chromatograph. Record the chromatograms and measure the response of the trehalose peak. Calculate the quantity in mg of trehalose in 1 ml of the sample solution

Triacetin   Transfer about 1 g of the sample, accurately weighed, into a suitable pressure bottle, add 25 ml of 1N potassium hydroxide and 15 ml of isopropanol, stopper the bottle, and wrap securely in a canvas bag. Heat in a water bath maintained at 98±2o for 1 h, allowing the water in the bath to just cover the liquid in the bottle. Remove the bottle from the bath, cool in airto room temperature, then loosen the bag, uncap the bottle to release any pressure, and remove the bag. Add 6 to 8 drops of phenolphthalein TS, and titrate the excess alkali with 0.5N sulfuric acid just to the disappearance of the pink colour. Perform a blank determination. Each ml of 0.5N sulfuric acid is equivalent to 36.37 mg of C9H14O6

Triammonium citrate   Dissolve about 3.5 g of the sample, accurately weighed, in 50 ml of water, add 50 ml of 1 N sodium hydroxide, boil for 15 min or until ammonia ceases to be evolved, add sufficient 1 N sulfuric acid to make the solution acid to phenolphthalein TS, boil for 5 min, cool, and titrate with 1 N sodium hydroxide, using phenolphthalein TS as an indicator. Each ml of 1 N sodium hydroxide is equivalent to 81.07 mg of C6H17N3O7.

Tricalcium phosphate	Weigh 200 mg of the sample to the nearest 0.1 mg and dissolve in amixture of 25 ml of water and 10 ml of dilute nitric acid TS. Filter, ifnecessary, wash any precipitate, add sufficient ammonia TS to the filtrate toproduce a slight precipitate, then dissolve the precipitate with the addition of 1 ml of dilute nitric acid TS. Adjust the temperature to about 50o, add 75 mlof ammonium molybdate TS, and maintain the temperature at about 50ofor30 min, stirring occasionally. Wash the precipitate once or twice with waterby decantation, using from 30 to 40 ml each time. Transfer the precipitate toa filter, and wash with a 1 in 100 potassium nitrate solution until the lastwashing is not acid to litmus paper. Transfer the precipitate and filter to theprecipitation vessel, add 40.0 ml of N sodium hydroxide, agitate until theprecipitate is dissolved, add 3 drops of phenolphthalein TS, and then titratethe excess alkali with N sulfuric acid. Each ml of N sodium hydroxidecorresponds to 6.743 mg of Ca3(PO4)2.
1,1,2-trichlorotrifluo roethane	Determine by Gas-liquid chromatography (see Volume 4):After determination of the total content of specified impurities, the balance consists of 1,1,2-trichlorotrifluoroethane together with any trace of other halogenated hydrocarbons that may be present. Calculate the percentage of 1,1,2-trichlorotrifluoroethane by the formula 100%-X, in which X is the percentage of other halogenated hydrocarbons determined as directed above.
Triethyl citrate	Weigh accurately about 1.5 g of the sample into a 500-ml flask equipped with a standard taper ground joint, and add 25 ml of isopropanol and 25 ml of water. Pipet 50 ml of 0.5 N sodium hydroxide into the mixture, add a few boiling chips, and attach a suitable watercooled condenser. Reflux for 1.5 h, then cool, wash down the condenser with about 20 ml of water, add 5 drops of bromothymol blue TS, and titrate the excess alkali with 0.5 N sulfuric acid. Perform a blank determination. Each ml of 0.5 N sulfuric acid is equivalent to 46.05 mg of C12H20O7.

Trimagnesium phosphate	Weigh to the nearest 0.1 mg, 200 mg of the ignited sample. Dissolve in a mixture of 25 ml of water and 10 ml of dilute nitric acid TS. Filter, if necessary, wash any precipitate, then dissolve the precipitate by the addition of 1 ml of dilute nitric acid TS. Adjust the temperature to about 50o, add 75 ml of ammonium molybdate TS, and maintain the temperature at about 50o for 30 min., stirring occasionally. Wash the precipitate once or twice with water by decantation, using from 30 to 40 ml each time. Transfer the precipitate to a filter, and wash with a 1 in 100 potassium nitrate solution until the last washing is not acid to litmus paper. Transfer the precipitate and filter to the precipitation vessel, add 40.0 ml of N sodium hydroxide, agitate until the precipitate is dissolved, add 3 drops of phenolphthalein TS and then titrate the excess alkali with N sulfuric acid. Each ml of N sodium hydroxide corresponds to 5.715 mg of $Mg_3(PO_4)_2$.
Trisodium citrate	Transfer about 350 mg of the sample, accurately weighed, to a 250-ml beaker. Add 100 ml of glacial acetic acid, stir until completely dissolved, and titrate with 0.1 N perchloric acid, using crystal violet TS as indicator. Perform a blank determination and make any necessary correction. Each ml of 0.1 N perchloric acid is equivalent to 8.602 mg of $C_6H_5Na_3O_7$.
Trisodium diphosphate	Using a previously dried sample, proceed as directed under Phosphate Determination as P_2O_5, Method 1, Inorganic components (Volume 4). Each ml of 1N sodium hydroxide consumed is equivalent to 3.088 mg of P_2O_5 or 5.307 mg of trisodium monohydrogen diphosphate on the dried basis.
Trisodium phosphate	Dissolve an accurately weighed quantity of the sample, equivalent to between 5.5 and 6 g of anhydrous Na_3PO_4, in 40 ml of water in a 400-ml beaker, and add 100 ml of 1 N hydrochloric acid. Pass a stream of carbon dioxide-free air, in fine bubbles, through the solution for 30 min to expel carbon dioxide, covering the beaker loosely to prevent loss by spraying. Wash the cover and

sides of the beaker with a few ml of water, and place the electrodes of a suitable pH meter in the solution. Titrate the solution with 1 N sodium hydroxide to the inflection point occurring at about pH 4, then calculate the volume (A) of 1 N hydrochloric acid consumed. Protect the solution from absorbing carbon dioxide from the air, and continue the titration with 1 N sodium hydroxide until the inflection point occurring at about pH 8.8 is reached. Calculate the volume (B) of 1 N sodium hydroxide consumed in the titration. When (A) is equal to, greater than, 2(B), each ml of the volume (B) of 1 N sodium hydroxide is equivalent to 163.9 mg of Na3PO4. When (A) is less than 2(B), each ml of the volume (A) - (B) of 1 N sodium hydroxide is equivalent to 163.9 mg of Na3PO4.

Urea   Accurately weigh about 0.5 g of the sample and dissolve in a 10% sulfuric acid solution and dilute to 100 ml with the same acid. Introduce 5.0 ml of this solution into a long-necked combustion flask, add 10 ml of sulfuric acid TS and heat gently until gas is no longer evolved. Boil gently for 10 min, cool and add cautiously 40 ml of water. Cool again and place in a steam-distillation apparatus. Add 50 ml of 10 N sodium hydroxide solution and distil immediately by passing steam through the mixture. Distil for 1 hour, collecting about 50 ml of distillate in 40 ml of a 4% w/v solution of boric acid. Add 0.25 ml of methyl red/methylene blue TS and titrate with 0.1 N hydrochloric acid. Carry out a blank determination. Each ml of 0.1 N hydrochloric acid is equivalent to 3.003 mg of CH_4N_2O.

Vegetable carbon   Measure total carbon in a dried sample by one of several methods or commercial instruments for carbon analysis, such as, instruments for C,H,O determinations or combustion/gravimetric carbon analysis. Dry the sample at 120 for 4 h before taking an accurately weighed analytical sample of an amount suitable for the specific method or instrument.

Xylitol

Determining using Liquid Chromatography: Internal standard solution: Transfer about 500 mg of erythritol, accurately weighed, into a 25 ml volumetric flask, dilute to volume with water, and mix. **Standard solution:** Transfer about 25 mg each of L-arabinitol, galactitol, mannitol, and sorbitol, accurately weighed, to a 100-ml volumetric flask, dilute to volume with water, and mix. To an accurately measured volume of this solution, add an accurately weighed amount of Reference Standard Xylitol (available from US Pharmacopeial Convention, Inc. 12601 Twinbrook Parkway, Rockville, MD 20852, USA) to obtain a solution with a known concentration of about 49 mg/ml. **Sample preparation:** Transfer about 5 g of the sample, accurately weighed, into a 100-ml volumetric flask, dilute to volume with water, and mix. **Chromatography:** Use a gas chromatograph equipped with a flame-ionization detector and a 2-m x 2-mm glass column packed with 3% liquid phase of 25% phenyl-25% cyanopropylmethylsilicone (OV-225 or equivalent) on silanized siliceous earth support (Chromosorb W-HP or equivalent). The carrier gas is nitrogen flowing at about 30 ml/min. The injector port temperature is 250o, the column temperature 200o and the detector temperature 250o. Chromatograph the derivatized Standards Solution prepared as directed under Procedure, and record the peak responses. The relative retention times corresponding to erythritol, L-arabinitol, xylitol, galactitol, mannitol, and sorbitol are usually about 1.0, 2.77, 3.90, 6.96, 7.63 and 8.43, respectively. The relative standard deviation of the response ratios of the derivatized Xylitol to the derivatized erythritol from three replicate injections does not exceed 2.0%. Procedure: Pipet 1 ml portions of the standards solution and the sample preparation into separate 100-ml, round-bottom boiling flasks. To each flask, add 1.0 ml of internal standard solution, and evaporate the respective mixtures to dryness on a water bath at 60o with the aid of a rotary evaporator. Dissolve each dry residue in 1 ml of pyridine, and add 1 ml of acetic anhydride to each flask. Boil each solution under reflux for 1 h to complete the acetylation.

Separately inject 1-μl portions of the derivatized solutions from the sample preparation and the standard solution into the gas chromatograph and measure the peak responses. Calculate the percentage of xylitol, on the as-is basis, by the formula: $100*Ws *RU/(WU*RS)$; where, WS = the weight, in mg, of Reference Standard Xylitol used for the Standard solution; WU = the weight, in mg, of the sample taken for the Assay preparation RU and RS = the ratios of peak responses of the derivatized analyte to the derivatized erythritol from the Internal Standard solution obtained from the Sample Preparation and the Standard Solution, respectively. Using the value obtained in the water determination, correct the percentage to the anhydrous basis.

Zeaxanthin (synthetic)

Substance	Relative retention time*	Approx. absolute retention time [min]
trans-zeaxanthin	1.00	17.7
cis isomers of zeaxanthin	1.38 – 1.46	24.4 – 25.8
12'-apo-zeaxanthinal	0.46	8.2
parasiloxanthin	0.96	17.0
diatoxanthin	1.16	20.5

The HPLC method of assay is designed to determine transzeaxanthin, the cis-isomers of zeaxanthins and zeaxanthin-related impurities: 12'-apo-zeaxanthinal, parasiloxanthin, and diatoxanthin. (NOTE: All solvents should be HPLC grade.). **Standards:** Trans-zeaxanthin, 12'-apo-zeaxanthinal, and diatoxanthin. (All trans-zeaxanthin, 12'-apo-zeaxanthinal, and diatoxanthin available from DSM Nutritional Products, Kaiseraugst, Switzerland. All-trans-zeaxanthin is also available from Fluka, Buchs, Switzerland). Solution 1: Accurately weigh 34 to 36 mg of 12'-apo-zeaxanthinal and transfer to a 100-ml volumetric flask. Add tetrahydrofuran to dissolve the substance and bring to volume. Solution 2: Accurately weigh 34 to 36 mg of diatoxanthin and transfer to a 100-ml volumetric flask. Add tetrahydrofuran to dissolve the substance and bring to volume. **Working standard:** Accurately weigh 69.0 to 71.0 mg of transzeaxanthin and transfer to a 100-ml volumetric flask. Add 50 ml of tetrahydrofuran, 1 ml of standard solution 1, and 1 ml of standard solution 2. Bring to volume with tetrahydrofuran. **Sample solution:** Accurately weigh 69.0 to 71.0 mg of sample and dissolve in 100 ml of tetrahydrofuran. **Mobile phase:** In a 2000-ml volumetric flask containing a small quantity of hexane, add 400 ml of ethyl acetate, 20 ml of 2-methoxyethanol, and 2.0 ml of N-ethyldiisopropylamine. Bring to volume with hexane. Chromatography apparatus and conditions: Column: Stainless steel; 250 x 4 mm; Column temperature: 25°; Stationary phase: Spherisorb Si, 3 μm or similar; Flow: Flow 1.0 ml/min; Detector: VIS 450 nm; Injection: 2.0 μl; Run time: 35 min; **Procedure:** Inject a 2.0 μl aliquot of the Working standard and measure the area of the peaks for trans-zeaxanthin, 12′-apo-zeaxanthinal, and diatoxanthin. Inject a 2.0 μl aliquot of the sample solution and measure the areas of the peaks for trans-zeaxanthin, cis-isomers of zeaxanthins, 12′-apo-zeaxanthinal, parasiloxanthin, and diatoxanthin.

www.ingramcontent.com/pod-product-compliance
Lightning Source LLC
Chambersburg PA
CBHW061509180526
45171CB00001B/106